SpringerBriefs in Information Systems

Series editor

Jörg Becker, Münster, Germany

For further volumes:
http://www.springer.com/series/10189

Pericles Antoniades

SOA Maturity Model

A Delphi-Derived Proposal for Inter-Enterprise Setups

Springer

Pericles Antoniades
Brighton
UK

ISSN 2192-4929 ISSN 2192-4937 (electronic)
ISBN 978-3-319-02452-3 ISBN 978-3-319-02453-0 (eBook)
DOI 10.1007/978-3-319-02453-0
Springer Cham Heidelberg New York Dordrecht London

Library of Congress Control Number: 2013953285

Printed on acid-free paper

Springer is part of Springer Science+Business Media (www.springer.com)

Preface

Service-Oriented Architecture (SOA) is the software architectural paradigm to which the bridging of the gap between IT and business is mainly attributed to, due to its capability and agility to easily and swiftly string together fairly large chunks of functionality to form ad-hoc applications that reflect existing or new business processes in order to address corresponding existing or new business requirements.

The Delphi technique was adopted as the vehicle for this work because its objective was to anticipate the adoption of SOA (and Web Services) in few years from the time this work was initiated and what might be the driving forces and prohibiting factors for the embracement of this new paradigm. It was due to this reason that the Delphi technique was considered to be the most appropriate methodology because it is used in cases when a forecasting about the state of or attitude toward a phenomenon is desired and the issue under investigation is very complex and requires the contribution of a pool of experts. Despite the criticism of Delphi's reliability, it is extensively used because it combines qualitative comments, which reflect insights of the group of participating members, with quantitative judgments.

In particular, a three-round questionnaire was distributed to participating IT experts with the objective to come up with a new SOA Maturity Model (MM) that would support inter-enterprise setups (beyond company's internal factors and domains). This would capture the model of contemporary organizations where it is a common practice that their core business processes are accomplished through digital "networks" that are spread not only within an entire organization, but also throughout its collaborating organizations.

Throughout this work, important "lessons" were learned that might be used as a guideline for future relevant ventures:

- Even though the Delphi technique proved to be beneficial when there is a need to obtain views and judgments from geographically dispersed knowledgeable people on the investigated issue, the fact that the panel of experts selected for this research originated from -only- the local industry and academia, did not affect its success and reliability because all the other factors (besides locality) were preserved: diversity of knowledge and expertise, acceptable group size whose members never functioned in a simultaneous face-to-face meeting, etc. However, it is advised to be evaluated, reviewed, and adjusted accordingly by academia and industry representatives elsewhere as well;

- The Delphi technique proved to be a potentially useful method for forecasting and obtaining experts' opinion on a topic. However, being that this was the first time this sort of method (variation of Delphi) was used for deriving a SOA MM, the possibility that shortcomings and limitations on the model might be observed is high. For instance, a reliable mechanism should be in place for inviting experts from all collaborating organizations, but at the same time preserving that they never function in a simultaneous face-to-face meeting;
- Delphi, by design, requests participating experts' opinion on an issue by providing possible responses, in the form of a series of questionnaires. No mechanism is in place for accepting critique on the questionnaires' design and content. It might be useful to let participants comment on topics where they are (proven to be) experts. This may lead to more complete and accurate final results;
- The electronic version of Delphi (known as "e-Delphi") might also be used in future works instead of the traditional version because it provides a number of benefits: quicker and easier implementation and management of the questionnaires, enhancement of the Delphi technique's reliability because no human intervention will be involved, etc.;

Finally, I would like to thank all colleagues, IT experts, who helped me with their feedback and insights to come up with this newly proposed SOA MM and, of course, my Editor Mr. Christian Rauscher, his Assistant Mrs. Barbara Bethke and the whole team at Springer whose contribution was invaluable for the publication of this work.

Special thanks go to my mother, Persa Antoniadou, who was my mother and father at the same time and who was always there. Thank you.

Contents

1 Introduction ... 1
References ... 2

2 The Problem and Research Questions ... 3
References ... 8

3 SOA, Maturity Models, SOA MM and Relevant Work ... 9
3.1 SOA ... 9
3.2 Maturity Models ... 11
3.3 SOA MM and Relevant Work ... 12
References ... 16

4 Research Approach and Methods ... 19
4.1 Delphi: Introduction, Implementation of the Method and Strengths and Limitations ... 19
4.1.1 Delphi: Introduction ... 19
4.1.2 Delphi: Implementation of the Method (The Process) ... 20
4.1.3 Delphi: Strengths and Limitations ... 20
4.2 Current Research Approach: A Variant to Delphi Technique ... 22
4.2.1 Applicability of the Approach ... 22
4.2.2 Delphi Study Structure ... 23
4.2.3 Codification and Experts Panel Teams Selection; Ethical Issues Addressed Through Standards, Restrictions and Rules ... 23
References ... 24

5 The Proposed SOA MM ... 27
5.1 Delphi Technique Applied: First-Round Questionnaire and Its Findings ... 27
5.2 Delphi Technique Applied: Second-Round Questionnaire and Its Findings ... 33

5.3 Delphi Technique Applied: Third-Round Questionnaire and Its Findings (Application/Use of the Proposed Inter-Enterprise SOA MM) 44

6 Proposed SOA MM Broadness and Applicability (in Particular: "Local" or "Global" SOA MM) 49
References 52

7 Conclusions, Limitations and Future Research 53
References 55

Appendices 57

Chapter 1
Introduction

Abstract Services, as perceived in the Service-Oriented Architecture (SOA), "are intrinsically un-associated units of functionality, which have no calls to each other embedded in them." (Wikipedia 2008). They typically implement functionalities most humans would recognize as a Service, "such as filling out an online application for an account, viewing an online bank statement, or placing an online booking or airline ticket order. Instead of Services embedding calls to each other in their source code, protocols are defined which describe how one or more Services can talk to each other." (Wikipedia 2008). This architecture (SOA) then relies on a business process expert to link and sequence services, in a process known as "orchestration", to meet a new or existing business requirement.

"The goal of SOA is to allow fairly large chunks of functionality to be strung together to form ad-hoc applications which are built almost entirely from existing software Services" (Wikipedia 2008) and its great promise is that the cost of creating any application should be negligible, as all of the software required already exists (in a repository of available Services); only "orchestration" is required to produce a new application.

The potential of SOA leads to the question of how advanced different organisations are in their SOA development and how can this be measured. Maturity is the means through which the capability and completeness in regards to SOA is being measured.

The notion of "maturity" has also been proposed for various management approaches (Quality Management, Process Management, Project Management, etc.) as a way to evaluate "the state of being complete, perfect, or ready" and the "fullness or perfection of growth or development" (Oxford University Press 2004).

This work aims at developing a new Capability Maturity Model (CMM)-based maturity model for the evaluation and scope of SOA initiatives using two distinctive features, in respect to similar works: to support inter-enterprise setups and the fact that it is evolved using a Delphi-variant technique. The proposed SOA

P. Antoniades, *SOA Maturity Model*, SpringerBriefs in Information Systems,
DOI: 10.1007/978-3-319-02453-0_1,

Maturity Model (MM) supports inter-enterprise setups (not only company's internal factors and domains), thus capturing the model of contemporary organizations where it is a common practice that their business processes are also spread throughout their collaborating organizations. The other differentiating feature of this work is the fact that a Delphi-variant technique is used to come up with the new maturity model and this is done because forecasting about SOA and its penetration in the subsequent years is desired and because for this prediction, the contribution of a pool of experts is required.

The newly proposed SOA MM is then applied to help the participating organizations position themselves in respect to (their current status towards) SOA (and guide them in achieving higher levels of SOA maturity) and anticipate their SOA maturity in five years time. Both of these objectives are achieved through the contribution of a panel of IT experts (as mandated by the Delphi technique).

The final objective of this research work is to come up with a conclusion about the "local" or "global" nature of the proposed SOA MM. Firstly, this will be checked with some of the panel experts and subsequently, by conducting a set of interviews with selected business owners and business practice managers.

The structure of this book is as follows: Chap. 2 defines the research problem and the associated research questions and Chap. 3 summarises relevant work already available. Following this, in Chap. 4, is an outline of the research approach and methods used throughout this work with a Delphi-variant technique being the "vehicle" for defining the new SOA MM. Chapter 5 details the newly proposed SOA MM and its application within the organizations of the panel experts indicating their current status with respect to SOA initiatives and what their SOA status will be in five years. Chapter 6 deals with the proposed SOA MM's breadth and applicability via interviews with some of the panel experts and selected business practitioners. Finally, Chap. 7 concludes with a review of the limitations of this research and suggestions for future relevant research.

References

Oxford English Dictionary. (2004). *The definitive record of the english language.*
Wikipedia. (2008). Retrieved September, 2008 from http://www.wikipedia.org.

Chapter 2
The Problem and Research Questions

Abstract: SOA, one of the latest developments in the Enterprise Architecture discipline, seems to be the development so desperately awaited for, because it enables Information Systems to flexibly adjust in order to meet potential changes in business requirements. Recognizing the potential and benefits provided by SOA, there should be a framework, a model, a SOA MM, that would help organizations position themselves in respect to their current SOA status and guide them to achieve higher levels of SOA maturity.

Keywords SOA • Enterprise architecture • Systems architecture • Loosely-coupled systems • Client–server computing • Distributed computing • Agility • SOA MM

Throughout the years, various attempts were undertaken to come up with such an enterprise architectural paradigm that would be thorough and flexible enough to satisfy the objective of IT: to address the current business needs and support the long-term business strategy of the organizations where IT (including hardware, software and networking infrastructure) has been adopted and used as a strategic asset.

SOA, being the latest of the evolutions in this area (Enterprise Architectures), seems to be the most promising enterprise architectural paradigm because it encompasses those properties (flexibility, agility, etc.) that enable the IT policy and business strategy to synchronize and the IT to facilitate the business in the most efficient manner.

It is due to those SOA's attributes (that is, the fact that SOA helps towards IT and business alignment and that IT is there to truly support business) that SOA has started to achieve acceptance in contemporary organizations which in turn, have started to measure their exposure and competence in it using corresponding maturity models (SOA MM).

None of the existing SOA MM, though, supports inter-enterprise setups and this is a significant deficiency because it is very common the practice to have inter-enterprise collaborations in the contemporary "networked" business environment.

P. Antoniades, *SOA Maturity Model*, SpringerBriefs in Information Systems,
DOI: 10.1007/978-3-319-02453-0_2, © The Author(s) 2014

This is one of the challenges of this research work along with the fact that the proposed SOA MM is derived using a Delphi-variant technique (considering that none of the existing SOA MM has been derived this way).

Into today's tough business environment, the reality is that the pressures do not lessen, the demands do not stabilize, and the Enterprise Architecture is lost in a myriad of tactical initiatives. As Paras (2004) and Zachman (1999) respectively argue: "The wild e-everything ride is over. Budgets are tighter and reality has set in. Executives tell us they must provide a solid, cost-effective IT foundation and simultaneously increase flexibility to respond to the increasingly diverse demands of the business. The effective use of information, technology, human resources, and investment capital must be balanced to achieve these goals. The solution is a portfolio focus, a return to disciplined, pragmatic approaches for strategy development and enterprise design, combined with robust processes for managing the enterprise portfolio of programs," (Paras 2004) and "Enterprise Architecture requires actual work. We keep looking for the 'quick fix,' a technological solution, a tool, a package, a new processor, the perennial 'silver bullet.' We wish we could simply throw money at the problem and have the pain go away," (Zachman 1999) but an evolutionary, standards-based approach to (enterprise) architecture needs to be devised to address the business needs today and support the long-term business strategy of contemporary organizations.

The architectural paradigm shifts observed during the last few years are depicted in Fig. 2.1. Looking at the figure, the first observation made is that there is a tendency to more "loosely coupled" systems. Moreover, it can be noticed that there is a trend towards more adaptable and flexible systems.

The mainframe systems of the 1960s were implemented as large blocks of functionality that ran on a single mainframe computer. On the contrary, services-oriented systems are implemented as discrete business services that are "loosely coupled" to other services running on -possibly- heterogeneous systems and platforms across an organization or beyond them.

Orton and Weick (1990) argue that three major definitions of "loosely coupling" are the dominant ones throughout the academic community. Glassman (1973) wrote that loose coupling is present when systems have either few variables in common or the variables they have in common are weak. Weick (1976) defined loose coupling as a situation in which elements are responsive, but retain evidence of separateness and identity. Later, he wrote that loose coupling is evident when elements affect each other "suddenly (rather than continuously), occasionally (rather than constantly), negligibly (rather than significantly), indirectly (rather than directly), and eventually (rather than immediately)" (Weick 1982).

The concept of "loosely coupled" is widely practiced in computing architectures. It is the foundation for the design of massively parallel computing systems.

This concept is also widely talked about, but far less widely practiced, in the software world. In some respects, the movement to three tier software architectures was a small step in the direction of loose coupling, at least at the level of standardizing interfaces across databases, application logic and presentation layers. In general, though, software has remained tightly coupled because of the

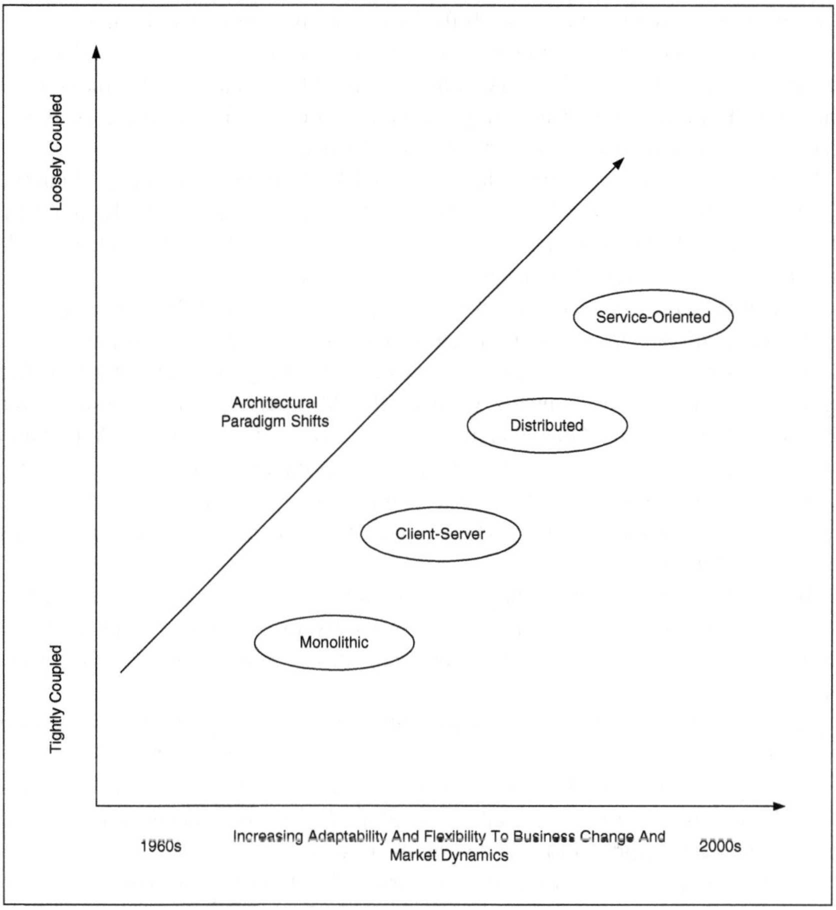

Fig. 2.1 Evolution of systems architecture [*Source* (Marks 2003)]

inability of major vendors to agree on a universal set of standards to define interfaces across software modules. However, this appears to be changing. Web services technology (a SOA implementation) is built upon a loosely coupled design philosophy (Hagel 2004).

Besides the tendency to becoming more "loosely coupled", the other shift that is observed in architectural paradigm throughout the period of the last forty years is the "migration" to more adaptable and flexible systems. Early mainframe systems used paper tape and punch cards to store data and programs. The use of mainframe processing time was strictly managed and allocated in sequential blocks or batches. On the contrary, services-oriented systems (implemented as discrete business services) are interconnected across an organization's computer network, where it is possible to locate and re-use services registered with a central registry of services.

Even though there have been shifts in the architectural paradigm towards a more "loosely coupled", flexible and easily adaptable model over the last 40 years, the current state of the IT industry is not yet an "ideal" place for business, especially for IT professionals (including the IT executives). For instance, today's IT executives have to deal with all sorts of technologies:

Business Applications—From large suites like Enterprise Resource Planning (ERP), Supply Chain Management (SCM), and Customer Relationship Management (CRM) packages to Portals and desktop productivity packages (like MS-Office, etc.), as well as industry-specific applications;

Distributed Computing Architectures—Client-Server architectures for database-centric applications and N-tier architectures for Internet or intranet purposes, etc.;

Middleware—The "glue" that keeps systems "talking" to each other, including message-oriented middleware (like IBM's MQ-Series), Application Servers (like BEA WebLogic, or Oracle's Application Server, etc.), Enterprise Application Integration (EAI) solutions, and transaction monitoring and processing systems;

Miscellaneous Systems—Mainframes, mid-range computers, servers of all types and sizes, desktop systems (PCs), and any number of special-purpose systems, depending on industry;

Bloomberg (2003b) argues that: "this plethora of technologies, while intended to *address* business issues, often *presents* issues that the IT executive must resolve. Most of these issues fall into three broad categories: *complexity, inflexibility,* and *brittleness.*"

All sorts of technologies encountered within contemporary organizations introduce unnecessary complexity. Today's enterprise IT environment contains many kinds of systems that work in many different ways. Enterprises must hire large, multi-skilled groups of professionals to develop, deploy, and manage the heterogeneous collection of applications and systems needed.

Besides complexity, the existence of various kinds of technologies introduces inflexibility, considering that almost all enterprises have existing business applications that are difficult to upgrade, interoperate with, and worst of all, impractical to replace. Furthermore, heterogeneous systems tend to be difficult to integrate, each exposing different interfaces with different rules. Integration is therefore an expensive, difficult process that yields inflexible distributed systems.

But, inflexible systems also encompass the risk of failure and are a source of instability and brittleness. Traditional approaches to building IT environments lead to really messy approaches to integration. As a result, when business processes or requirements change, IT departments must either undertake expensive, risky upgrade projects, or simply drop the existing applications and systems because they no longer meet the business requirements.

Of course, the problems of complexity, inflexibility, and brittleness are nothing new in the enterprise. The need for a new solution (or approach) that will address all these issues encountered in the IT landscape today, seems to be, more than ever, an imperative matter. Today's IT executives need fresh approaches in dealing with heterogeneous environments and an increasing pace of change, in the face of tight budgets and a tough economy.

As a response to that need, a new approach to IT resources integration and systems architecture is gaining traction in enterprises across many industries. Known as *Service-Oriented Architecture* (SOA), this new way of thinking about how to integrate IT resources and access applications functionality in the enterprise aims to address most of the issues faced by enterprises today. "SOAs have the potential to rise to the challenges of brittle application infrastructures, inflexible technology, and high-risk, high-cost IT. Fundamentally, SOAs have the flexibility and responsiveness to finally enable business priorities to drive technology decisions. On the other hand, building service-oriented infrastructures is not easy. It requires commitment and expertise. The long-term business benefits of SOAs, however, can justify such investments. Many enterprises have already implemented SOAs and achieved quantifiable benefits from their investment in this new architectural approach." (Bloomberg 2003a).

As a consequence, a SOA MM might be beneficial because it could help organizations position themselves in respect to SOA and guide them to achieve higher levels of SOA maturity. In particular, a maturity model is a framework that describes, for a specific area of interest (SOA in our case), a number of levels of sophistication at which activities in this area can be carried out. Existing literature (and SOA MM proposed by IT and other companies in the market) indicates important criteria to judge the level of SOA maturity within a single enterprise. However, identifying such criteria in an inter-enterprise environment is hardly addressed at all. Moreover, no inter-enterprise SOA MM has been defined for the time being.

According to resource dependence theory, a theory formulated in the 1970s by Pfeffer and Salancik (1978), "the elemental structural characteristics of environments are concentration, the extent to which power and authority in the environment are widely dispersed; munificence, or the availability or scarcity of critical resources; and interconnectedness, the number and pattern of linkages, or connections, among organizations. These structural characteristics, in turn, determine the relationships among social actors—specifically, the degree of conflict and interdependence present in the social system. Conflict and interdependence, in turn, determine the uncertainty the organization confronts." This theory clearly states that in an inter-enterprise environment, organizations manage their dependence with the goal of decreasing uncertainty by creating formal inter-enterprise structures that formalize their relations with other organizations. In such a condition, organizations begin to collaborate together for a common purpose.

Therefore, the objective of this research is to bridge this gap; that is, to propose an inter-enterprise SOA MM by addressing a set of research questions:

- How many levels, stages of maturity will the proposed SOA MM contain?
- What are the domains (and the relevant focus areas) that will be included in the proposed SOA MM?
- Why is the chosen research methodology (a Delphi-variant technique) considered to be appropriate for this research?
- What are those features that need to be included in the proposed SOA MM to make it inter-enterprise (and that are not found in the existing SOA MMs)?

Besides the major research questions above, the following ones are also discussed throughout this research:

- What is the relationship of the proposed SOA MM to existing ones and to the CMMI (CMM, Integration-specific) model?
- How do we know that the proposed SOA MM is valid and appropriate for inter-enterprise settings?
- What potential benefits and problems does the inter-enterprise nature of the proposed SOA MM bring?

References

Bloomberg, J. (2003a). *Service-oriented architecture: Why and how?* (p. 4). Massachusetts.

Bloomberg, J. (2003b). *Service-oriented architecture: Why and how?* (p. 5). Massachusetts.

Glassman, R. B. (1973). Persistence and loose coupling in living systems. *Behavioral Science, 18*, 83–98.

Hagel, J. (2004). Retrieved January, 2004, from http://www.johnhagel.com/blog20021009.html

Marks, E., et al. (2003). *Executive's guide to web services*. New Jersey: Wiley and Sons Inc.

Orton, J. D., & Weick, K. E. (1990). Loosely coupled systems: A reconceptualisation. *Academy of Management Review, 15*(2), 203–223.

Paras, G. (2004). METAGroup Vice President and Director of Enterprise Planning an Architecture Strategy. Retrieved January, 2004, from (http://www.meta.com)

Pfeffer, J., & Salancik, G. R. (1978). *The external control of organizations: A resource dependence perspective*. New York: Harper & Row.

Weick, K. E. (1976). Educational organizations as loosely coupled systems. *Administrative Science Quarterly, 21*, 1–19.

Weick, K. E. (1982). Management of organizational change among loosely coupled elements, In P. S. Goodman & Associates (Eds.), *Change in organizations* (pp. 375–408). San Francisco: Jossey-Bass.

Zachman, J. (1999). *Enterprise architecture: The past and the future*. DM Review.

Chapter 3
SOA, Maturity Models, SOA MM and Relevant Work

Abstract SOA, based on the loosely coupled principle of systems architecture, can address the complexity, inflexibility, and brittleness issues of existing approaches to integration through reusable services that can be easily and flexibly "orchestrated" into different business processes to meet new or existing business system requirements. Carnegie Mellon University's Software Engineering Institute proposed the Capability Maturity Model (CMM) which became the prototype based on which all other maturity models evolved. CMM helps toward evaluating the capabilities and "maturity" of an organization in regards to its software development processes. The Model has then been varied and extended to be applied to IT Infrastructure Management, Enterprise Architecture Management, Knowledge Management, etc. CMMI (CMM, Integration-specific) is one such a maturity model which was evolved from the CMM. CMMI helps integrate traditionally separate organizational functions, set process improvement goals and priorities, provide guidance for quality processes and a point of reference for appraising current processes. Based on that all other SOA MM were emerged.

Keywords SOA • Service provider • Publish • Service registry • Find • Service requestor • Bind • Maturity models • SOA MM • SOA MM levels

3.1 SOA

SOA is an approach to enterprise business systems and applications that considers software resources as services available and discoverable on a network. Such services provide functionality to the business while hiding the underlying implementation details. SOA consists of three components and there are three major operations through which the SOA participants interact (Fig. 3.1).

The "Service Provider" creates and publishes the services in the "Service Registry" which is responsible for registering and categorizing the services. These services are being "consumed" by the "Service Requestor" which discovers the

P. Antoniades, *SOA Maturity Model*, SpringerBriefs in Information Systems,
DOI: 10.1007/978-3-319-02453-0_3, © The Author(s) 2014

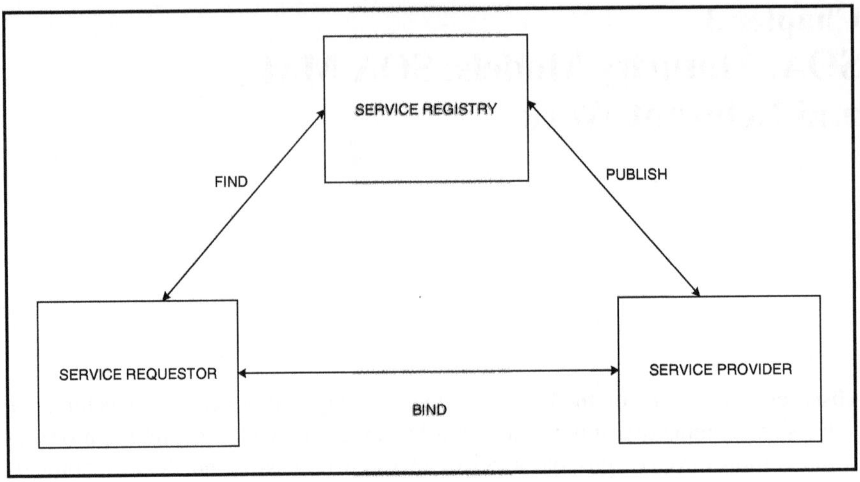

Fig. 3.1 The service oriented architecture [*Source* (Cutlip 2001)]

required services by searching the registry and then invoking them by communicating with the actual "Service Provider".

The interaction between the major components of the SOA architecture is done via means of three basic operations: *Publish* (the "Service Provider" publishes its services to the "Service Registry"), *Find* (the "Service Requestor" locates, searches, and discovers the services stored by the "Service Registry"), and *Bind* [the "Service Requestor" binds and uses the service(s) provided by the "Service Provider" (through the "Service Registry")].

SOA is primarily differentiated from previous architectural paradigms (Monolithic, Client-Server, and Distributed) due to certain distinct characteristics: the fact that the exchange of information between the communicating parties ("Service Provider" and "Service Requestor") is realized using "Open" interface standards (particularly, XML-based technologies), the fact that in SOA we strive to create coarse-grained services, meaning that each service delivers a significant piece of data or functionality which represents a well-defined step in a business process or transaction. On the contrary, in the applications or systems that are based on the Monolithic, Client-Server, or Distributed architectural paradigms, we create fine-grained components that implement low-level technical functions (and not real business functionality).

The most decisive factor, though, that distinguishes a SOA-based system from systems that are based on the other architectural paradigms, is its "loosely coupled" attribute. "Loosely coupled is an attribute of systems, referring to an approach to designing interfaces across modules to reduce the interdependencies across modules or components—in particular, reducing the risk that changes within one module will create unanticipated changes within other modules. This approach specifically seeks to increase flexibility in adding modules, replacing modules and changing operations within individual modules" (Hagel 2004).

Indeed, traditional systems are tightly coupled (which means that the implementation of a "Service Provider" had to be closely tied to the implementation of the "Service Requestor" for the communicating parties to "talk" to one another), as opposed to SOA-based systems where the exchange of information between communicating parties takes the form of exchanging flexible, loosely tied services.

Finally, the other attribute that makes SOA distinct is the fact that in a SOA-based environment, services are published in the "Service Registry". "Service Registry" is a central repository, either within an organization or on the Web, which can be used to *dynamically discover* services at run-time, as opposed to rigid, monolithic and very specific references (to components) observed in the traditional architectural paradigms.

To conclude, SOA could address the complexity, inflexibility, and brittleness issues of existing approaches to integration through coarse-grained, loosely coupled, reusable services that could be easily and agilely "orchestrated" into different business processes to meet new or existing business system requirements every time.

3.2 Maturity Models

The concept of "maturity" literally means the change from one initial state to another, more advanced state (Fraser et al. 2002). Implicitly, the concept of "maturity" incorporates the notion of gradual evolution through intermediate states. The evolutionary nature of these states is expressed by Shapiro (1996) pointing out the fact that all companies that improve their processes pass through phases of evolution.

Maturity as a measure to evaluate the capabilities of an organization in regards to a certain discipline has become popular since the Capability Maturity Model (CMM) which was proposed by the Software Engineering Institute at the Carnegie Mellon University (Paulk et al. 1993). Whilst the original CMM focuses specifically on the evaluation of software development processes, the model has been varied and extended and is now applied to IT Infrastructure Management, Enterprise Architecture Management and Knowledge Management to name a few.

The roots of maturity approaches, though, lie in quality management. Crosby (1979) introduced the first maturity model named "Quality Management Maturity Grid." Since that first maturity model, many models have been introduced in different areas. Maturity is not only related to quality management; there are maturity models that measure the capacity and maturity of other disciplines and practices, too. For instance, there are maturity models that measure the maturity of a project management practice (Kerzner 2005), of software maintenance (April et al. 2005), of business processes (Rosemann and de Bruin 2005) and of outsourcing (Adelakun 2004). Maturity models are also able to explain the maturity of an enterprise architecture by describing the ability to manage its development, implementation and maintenance (van der Raadt et al. 2005).

Alleman (2005) argues that the idea of maturity assessment aims at addressing three major aspects in an organization: identifying risks, focusing improvement and identifying areas from which it can have financial and other benefits. So, it can be said that the establishment of an organization's maturity can be the first step toward the development of its improvement strategy which in turn, can help it establish goals for process improvement and identify opportunities for business optimization.

A maturity model can also allow an organization to improve its processes by benchmarking itself against other organizations. By benchmarking its processes, an organization can measure the efficiency of its operations against similar operations of other entities.

Nowadays, many organizations are trying to setup and execute a process management improvement strategy in an attempt to gain real financial and other benefits. Unfortunately, as Gainer (1998) notes, there is often significant resistance in some organizations to formalized process initiatives. In particular, some organizations are so preoccupied with market and daily demands that a serious process improvement initiative is practically impossible. Thus, seeking formal improvement within those organizations can be time consuming and expensive, but the rewards in the quality of the processes can be significant.

Moreover, a maturity model can offer an improvement path to the firm in terms of its predictability which is realized through a greater accuracy in anticipating results and thus effectively contributing to the planning and budgeting of resources allocation and utilization (Garcia Romero 2001).

An additional benefit that can be provided by a maturity model is the fact that it can allow a possible interested party to be certain of the potential of an organization. Depending on the level of maturity that an organization has, an external party (another organization, a stakeholder, a client, or a supplier) can have a clearer idea about the organization's capabilities and potential, in order to decide whether it is risky or not to have economic dealings with it.

Summarizing, with maturity models the organizations can determine their maturity and capability in various areas, and this might enable them to evolve toward a culture of process improvement excellence which, in turn, leads to greater efficiency of operations, more accurate planning, safer decision making, less risks and higher credibility.

3.3 SOA MM and Relevant Work

Recently a number of models to measure the maturity of SOA initiatives by organizations has been proposed. The common base for the majority of these models has been the CMM. This model is "a process improvement approach that provides organizations with the essential elements of effective processes. It can be used to guide process improvement across a project, a division, or an entire organization. CMMI (CMM, Integration-specific) helps integrate traditionally separate

Table 3.1 CMMI model (Based on: CMMI 2006)

CMMI level	CMMI level brief description
Level 1: Performed	Base practices are performed in order to achieve specific goals;
Level 2: Managed	A managed process institutionalized by, for example, establishing policy, creating standards and training people;
Level 3: Defined	The process should be institutionalized and tailored to the company to be able to apply it to the whole company;
Level 4: Quantitatively managed	The process should be controlled by quantitative techniques;
Level 5: Optimizing	This level is characterized by a process adopted to meet relevant current and projected business objectives through continuous process improvement;

organizational functions, set process improvement goals and priorities, provide guidance for quality processes and a point of reference for appraising current processes." (CMMI 2006).

Pyster (2005) describes CMMI model as a relatively complete and widely used framework for process improvement in software development. CMMI's maturity levels are defined as evolutionary plateaus of process improvement, which should help to predict the future performance of an organization by describing the range of expected results. In particular, five levels of process improvement maturity are defined, and the level an organization is currently located (in terms of its process improvement stage) is determined according to the fulfillment (or not) of well-(pre-)defined indicators (Table 3.1).

According to CMMI (2006), both business and technical benefits are expected from the adoption of the CMMI framework by organizations. Among others, these include: "the substantial reduction in systems integration and test time with greater probability of success", "CMMI is considered to be the cause of integration of, and interaction among, the various engineering functions", "CMMI contribute to the overall project and organization success", "CMMI leverages previous process improvement investments" and the "increased focus and consistency in requirements development and management, systems design and development, systems integration, risk management, and measurement and analysis." (CMMI 2006).

Bush and Dunaway (2005) also argue that CMMI emphasizes the importance of detecting defects early in process improvement initiatives and subsequently preventing their occurrence: "the process improvement initiated by this model involves an organizational discipline that recognizes and deals with problems early, accepts independent quality reviews, and promotes discomfort when quality procedures are missing which in turn motivates positive change." (Bush and Dunaway 2005).

Furthermore, the real-world benefits enjoyed by organizations whose "software development"-related projects were implemented using the CMMI model proved the model's potential. Indicatively, "Lockheed Martin", a military and defense systems company, managed to increase software productivity by 30 % and decrease defect finding and fixing costs by 15 %; for "General Motors Corporation", an automotive manufacturing company, the use of CMMI model resulted in improved projects' schedules, that is, projects met milestones and were fewer delays; CMMI helped "Thales ATM", an air traffic management company, to see into the future with a good level of confidence, have an increasing number of processes under statistical control and achieve return on investment due to earlier defect detection, improved risk management and better control of projects; Improvements caused by CMMI adoption were also realized by "Bosch Gasoline Systems", a leading global supplier of technology in the automotive industry: the internal on-time delivery improved by 15 %, the first pass yield improved by 10 % and there was a reduction in error cases in the factory by one order of magnitude; "J.P. Morgan Chase and Co.", a leading global financial services firm, also enjoyed CMMI-related benefits: there was an improved predictability of delivery schedule, a reduction of post-release defects, a reduced severity of post-release defects and an increased throughput.

Besides the real-world cases and benefits enjoyed by organizations that adopted CMMI, a set of CMMI's strengths and weaknesses, separated in five major domains (Size, Structure, Scope, Interpretability, Sustainability of Commitment), testified by engineers of a leading Systems Integrator, EDS (2007), which plans, designs and implements CMMI-based projects for other organizations, constitute a good indicator for the model's potential and shortcomings mainly because it is based on their extensive exposure and experience in the area.

As far as the size of the model is concerned, EDS engineers argue that even though the model provides a large amount of background and guidance information based on industry best practices which can be valuable if used correctly, the size of the model can result in information overload, leading to either "paralysis" or "blind practice implementation." (EDS 2007).

Regarding structure, the strengths identified are the facts that: "by providing a hierarchy of process areas, CMMI promotes an evolutionary approach to improvement which is more effective and sustainable" and "the separation of process areas allows for more focused and manageable process improvement efforts." (EDS 2007). On the contrary, the large number of separate process areas makes it difficult to clearly identify and understand their relationships and the fact that in some cases, organizations develop process architectures which reflect CMMI's process areas rather than their business operations, are the major weaknesses of the CMMI model in respect to its structure.

CMMI's limitation of scope allows for a greater level of detail and guidance for project-related processes. Its limited scope, though, provides limited support for non-project related business functions. According to EDS (2007), "CMMI goals and generic practices are worded in such a way as to be applicable to a wide range of organizations and projects" and "the CMMI contains a large amount of interpretive guidance for organizations and projects focused on large-scale development" are regarded as CMMI's strong points, in respect to interpretability. On the other hand, the fact that the CMMI model refers to specific practices and the interpretive guidance tends to be worded to apply to large-scale development and limited guidance is provided to interpret this for "small enhancements" or "non-development" projects is considered to be the major shortcoming as far as the interpretability of the CMMI model is concerned.

Finally, in respect to the CMMI's sustainability of commitment domain, EDS engineers argue that the fact that CMMI provides maturity levels as a means for setting and evaluating progress towards process improvement targets and that it has a strong focus on measurement which assists with determining the return on investment for process improvement activities, are among the strengths of CMMI regarding this domain. The weaknesses, though, of the CMMI model, in respect to the sustainability of commitment domain, according to the EDS (2007), constitute the facts that the "implications of targets are not always understood and this can lead to the CMMI becoming the end in itself and not a means to drive improvements" and that the "timeframes for achievement of CMMI targets may be inconsistent with short-term business cycles." (EDS 2007).

To conclude, the CMMI model incorporates both strong and weak attributes. Even though one may argue that: (a) CMMI's underlying assumption: "good processes lead to good results" is both a questionable assumption and one that leads to too much introspection about process and too little focus on actual results, and (b) it is impossible, especially for any non-trivial project, to get one level of the CMMI model perfected before moving on to the next phases (as it also happens to all staged models), CMMI remains a widely accepted and a relatively complete model that has been designed for process improvement in software development. It is due to all these CMMI's properties and strengths that most of the SOA maturity models are having CMMI as a common base.

Currently, SOA is the new trend in IT Architecture and consequently specialized maturity models (among others, IBM 2006; HP 2006; Oracle 2006, Sonic Software Corp. et al. 2005; Arsanjani and Holley 2005) were created to describe the maturity of a SOA initiative. The SOA Maturity Model might become the enabling force for an enterprise to be drawn into the "vehicle" of this new software paradigm (SOA) because it designates the degree to which an enterprise might benefit from the SOA adoption at the point in time the model is being applied and what actions are needed to proceed to higher levels in the SOA maturity scale and thus further "reap" the value "encapsulated" within SOA. It is due to this SOA property that the IT giants (IBM, Accenture, Microsoft, Sun Microsystems, Oracle, HP, BEA, etc.) and others (for instance, CBDI Forum) not only came up with formal guidelines (or even planning tools) for successfully adopting SOA, but also started proposing SOA maturity models. A detailed presentation of these SOA maturity models can be found in Appendix I (titled: "Major SOA Maturity Models").

All these SOA MM, though, lack an important ingredient that this research tries to address: consideration of domains (and corresponding focus areas) observed in inter-enterprise environments. I think that this is very important as changes in the business environment and competition force companies to rethink the way they are doing business. More and more organizations nowadays take advantage of the next level of re-engineering approaches which capitalize on connecting and aligning one company's operations with other companies to meet important organizational goals. The origin of these interconnected, inter-enterprise business structures, called networked businesses, are explained by the resource dependence theory (Pfeffer and Salancik 1978). In summary, the resource dependence theory tells us that organizations must study themselves in relation to the organizations with which they want to share resources. In such a study, organizations need to give special attention to external control which they could face when their processes depend partially, or completely, on other organizations' resources. Based on this, it is considered that the proposed SOA MM should include inter-enterprise parameters to be complete and valuable to the organizations that will adopt it.

References

Adelakun, O. (2004). It outsourcing maturity model. In L. T, S. T & K. S, (Eds.), *Proceedings of the 13th European Conference on Information Systems, The European IS Profession in the Global Networking Environment, ECIS 2004*, June 14–16, 2004. Turku, Finland, DBLP (2004).

Alleman, G. (2005). Some thoughts on project management organizational maturity models, Project management forum.

April, A., Hayes, J. H., Abran, A., & Dumke, R. R. (2005). Software maintenance maturity model (SM, mm): The software maintenance process model. *Journal of Software Maintenance, 17*(3), 197–223.

Arsanjani, A., & Holley, K. (2006). Increase flexibility with the service integration maturity model (simm). Retrieved March, 2006, from http://www-128.ibm.com/developerworks/web services/library/ws-soa-simm/

Bush, M., & Dunaway, D. (2005). *CMMI assessments: Motivating positive change*. New Jersy: Addison-Wesley Professional.

CMMI (2006). Retrieved December, 2006, from http://www.sei.cmu.edu/cmmi/

Crosby, P. B. (1979). *Quality is free*. New York: McGraw-Hill.

Cutlip, R. (2001) Web services: The new web paradigm. *DB2 Magazine*, Quarter (4).

EDS (2007). Retrieved November, 2007, from http://www.eds.com

Fraser, P., Moultrie, J. & Gregory, M. (2002). The use of maturity models/grids as a tool in assessing product development capability. IEEE International Engineering Management Conference, Cambridge, August 19–20.

Gainer, J. (1998). Process improvement: The capability maturity model, IT management web.

Garcia Romero, C. I. (2001). El Modelo de Capacidad de Madurez y su Aplicaci' on en Empresas Mexicanas de Software (BA thesis, Universidad de las Am 'ericas, Cholula, Puebla, Mexico, 2001).

Hagel (2004). Retrieved January, 2004, from http://www.johnhagel.com/blog20021009.html

HP. (2006). Retrieved December, 2006, from http://www.hp.com

IBM. (2006). Retrieved December, 2006, from http://www.ibm.com

Kerzner, H. (2005). *Using the project management maturity model : Strategic planning for project management.* New York: Wiley.

Oracle. (2006). Retrieved December, 2006, from http://www.oracle.com

Paulk, M. C., et al. (1993). *The capability maturity model for software.* Version 1.1 (No. CMU/SEI-93-TR-24): Software Engineering Institute.

Pfeffer, J., & Salancik, G. R. (1978). *The external control of organizations: A resource dependence perspective.* New York: Harper & Row.

Pyster, A. B. (2005). What beyond CMMI is needed to help assure program and project success? In M. Li, B. W. Boehm & L. J. Osterweil (Eds.), Revised Selected Papers of the International Software Process Workshop, SPW 2005 (Vol. 3840, pp. 75–82). Beijing, China: Lecture Notes in Computer Science, Springer.

Rosemann, M., & de Bruin, T. (2005). Towards a business process management maturity model. In B. D, R. F, K. J, A. D, W. R, E.-D. P, B. J, B. F &W. C (Eds.), Proceedings of the 13th European Conference on Information Systems, Information Systems in a Rapidly Changing Economy, ECIS 2005. Regensburg, Germany, May 26-28, DBLP, 2005.

Shapiro, R. (1996). Stages in the evolution of the product development process. In M. E. McGrath (Ed.), *Setting the PACE in product development: A guide to product and cycle time excellence*. Boston, MA: Butterworth-Heinemann.

Sonic Software Corp., AmberPoint Inc., BearingPoint Inc. & Systinet Corp. (2006). A new service-oriented architecture (soa) maturity model. Retrieved March, 2006, from http://www.systinet.com/dl/SOA-Maturity-WP.pdf

van der Raadt, B., Hoorn, J. F. & van Vliet, H. (2005). Alignment and maturity are siblings in architecture assessment. In O. Pastor & J. F. e Cunha (Eds.), Proceedings of the 17th International Conference, CAiSE 2005 (Vol. 3520, pp. 357–371). Porto, Portugal: Lecture Notes in Computer Science, Springer, June 13–17, 2005.

Pyster, A. B. (2005). What beyond CMMI is needed to help assure program and project success? In M. Li, B. W. Boehm & L. J. Osterweil (Eds.), Revised Selected Papers of the International Software Process Workshop, SPW 2005 (Vol. 3840, pp. 75–82). Beijing, China: Lecture Notes in Computer Science, Springer.
Rosemann, M., & de Bruin, T. (2005). Towards a business process management maturity model. In B. D., R. F., K. J., A. D., W. R., E.-D. P., B. J., B. F., & W. C. (Eds.), Proceedings of the 13th European Conference on Information Systems, Information Systems in a Rapidly Changing Economy, ECIS 2005, Regensburg, Germany, May 26–28, DBLP, 2005.
Shapiro, S. (1996). Stages in the evolution of the product development process. In M. E. McGrath (Ed.), Setting the PACE in product development: A guide to product and cycle-time excellence. Boston, MA: Butterworth-Heinemann.
Sonic Software Corp., AmberPoint Inc., BearingPoint Inc., & Systinet Corp. (2006). A new service-oriented architecture (SOA) maturity model. Retrieved March, 2009, from http://www.sys-con.com/SOA Maturity WP.pdf.
Van der Raadt, B., Hoorn, J. F., & van Vliet, H. (2005). Alignment and maturity are siblings in architecture assessment. In O. Pastor & J. Falcão e Cunha (Eds.), Proceedings of the 17th International Conference, CAiSE 2005 (Vol. 3520, pp. 357–371). Porto, Portugal: Lecture Notes in Computer Science, Springer. June 13–17, 2005.

Chapter 4
Research Approach and Methods

Abstract The Delphi method, which was chosen to be the "vehicle" of this work, is used for deriving experts' opinion on a topic and generating forecasts in technology, education, and other areas. Its objective is fulfilled by reaching a consensus by a selected group of experts whose opinion on an issue is gathered via a series of consecutive questionnaires. In particular, a Delphi-variant technique has been chosen to be the most appropriate methodology for this work because of the availability of a pool of local SOA experts that contributed toward the development of a new SOA MM and the innovativeness and potential provided by Delphi itself.

Keywords Delphi • Panel of experts on SOA • Questionnaires • Consensus

4.1 Delphi: Introduction, Implementation of the Method and Strengths and Limitations

4.1.1 Delphi: Introduction

The Delphi method is a structured method for deriving experts' opinion on a topic (Adler and Ziglio 1996) and generating forecasts in technology, education, and other fields (Cornish 1977). Its objective is fulfilled by reaching a consensus and position by a selected group of experts whose opinion on a defined issue is gathered via a series of questionnaires (Helmer 1983).

Even though the experts' judgment on a defined issue might be considered to be subjective, it is believed to be more reliable than individual's statements and is thus more objective in its outcome (Johnson and King 1988; Masini 1993). As Linstone and Turoff (1975) also argue: "Delphi may be characterized as a method for structuring a group communication process, so that the process is effective in allowing a group of individuals, as a whole, to deal with a complex problem."

P. Antoniades, *SOA Maturity Model*, SpringerBriefs in Information Systems,
DOI: 10.1007/978-3-319-02453-0_4,

The Delphi method, which has its origins back in 1953 and it was created by Olaf Helmer and Norman Dalkey at the RAND corporation to address future military issues, has been widely accepted because of its flexibility and applicability in various areas: it is a method for structuring a group communication process to facilitate group problem solving (Linstone and Turloff 1975), it can be used as a judgment, decision-aiding or forecasting tool (Rowe and Wright 1999), and can be applied to program planning and administration (Delbeq et al. 1975). It can also be used when there is incomplete knowledge about a problem or phenomenon (Adler and Ziglio 1996; Delbeq et al. 1975). Moreover, the Delphi method can be used to investigate what does not yet exist (Czinkota and Ronkainen 1997; Halal et al. 1997; Skulmoski and Hartman 2002).

4.1.2 Delphi: Implementation of the Method (The Process)

According to Fowles (1978), the Delphi method involves: (a) selection of panel of expert based on pre-determined criteria (for example, proven academic qualifications and professional expertise on a defined issue under investigation, publications, etc.); (b) formulation of the first-round questionnaire which is distributed to the selected experts; (c) completion of the questionnaire (by the panel experts) and its submission to the researcher (or codification team) who (or which) is performing the study; (d) analysis of the participating experts' answers and identification of the general tendency in their responses on the various points raised; (e) check whether appearance of convergence between the participating experts' points of view is observed (this normally starts appearing after the second-round questionnaire); if not, a new questionnaire is prepared and sent out to the panel experts; this cycle (constituting of stages –c- to –e-) is repeated until a sufficient convergence of opinions is reached; (f) preparation of a summary of the process followed and drawing up of the final report with findings and conclusions (when, of course, the expected convergence of experts' opinions is reached); These main stages of a typical Delphi method process implementation are shown in Fig. 4.1.

4.1.3 Delphi: Strengths and Limitations

Throughout the years, various criticisms have been leveled at Delphi. In particular, according to Gordon and Hayward (1968), Gatewood and Gatewood (1983), and Adler and Ziglio (1996), a major problem identified into the implementation and application of Delphi surveys has been the experts' approach toward the investigated issue. In particular, most of the times, experts treat the issues under investigation in a very simplistic way, failing to have a more holistic view and overlooking all involved factors and parameters.

Other criticisms of Delphi studies include the works by Erffmeyer et al. (1986), Schmidt (1997), Turoff (1970), and van de Ven and Delbecq (1974) who argue

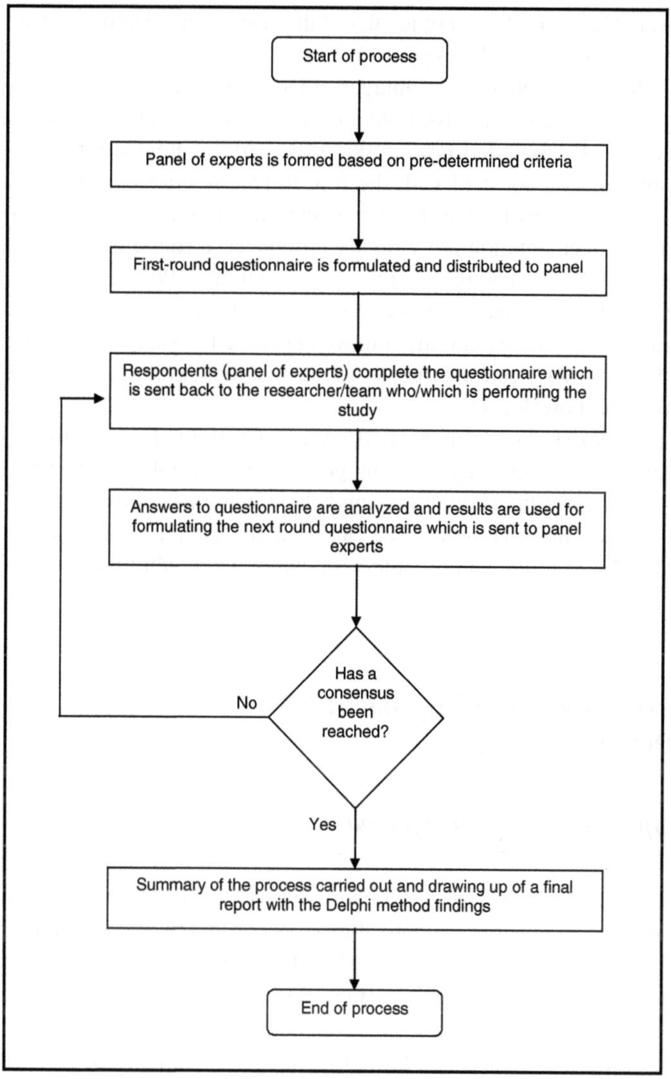

Fig. 4.1 Typical Delphi method process implementation

that Delphi study design is not so flexible, and the conclusions drawn by Ono and Wedemeyer (1994) and Woudenberg (1991) who support that the accuracy and validity of Delphi study outcomes is something that is questionable.

Also, according to Makridakis and Wheelwright (1978), the key problems reported, with regards to the Delphi method implementation, include: "poor internal consistency and reliability of judgments among experts, and therefore low reproduce-ability of forecasts based on the results elicited; sensitivity of results to ambiguity and respondent reactivity in the questionnaires used for data collection; difficulty in assessing the degree of expertise held by participating experts."

Moreover, Martino (1978) argues that: "the results of a Delphi survey are only as valid as the opinions of the experts involved."

On the other hand, reported advantages of Delphi studies include: (a) the fact that anonymity leads to more creative outcomes and adds richness to data (van de Ven and Delbecq 1974; Okoli and Pawlowski 2004), (b) the argument that the possibility of having issues associated with the face-to-face interactions (that is, dominate personalities, conflict and group pressures, etc.) as the ones occurred throughout a Delphi study implementation are virtually eliminated through the rules and restrictions imposed while carrying out the study (Helmer 1967; Loo 2002; Murphy et al. 1998) and (c) the fact that geographic boundaries and associated travel and co-ordination factors are essentially removed because the study can be implemented either via fax or other electronic means -electronic mail, etc.- (Loo 2002; Okoli and Pawlowski 2004; Powell 2003).

There are also studies (Ament 1970; Wissema 1982; Helmer 1983; Ono and Wedemeyer 1994) suggesting that, in general, the Delphi method is useful in the cases of single-dimension issues and their outcomes proven to be accurate. Similarly, Ascher and Overholt (1983) argue that: "Delphi studies have an excellent record of forecasting, among others, in computer capability advances."

4.2 Current Research Approach: A Variant to Delphi Technique

4.2.1 Applicability of the Approach

It is interesting to look at what researchers -who extensively work with the Delphi method- suggest of Delphi's applicability: according to Okoli and Pawlowski (2004) and Ono and Wedemeyer (1994), Delphi studies are considered beneficial when one deals with complex issues; Bass (1983) argues that Delphi can be used when one seeks to combine views to improve decision making; Delbecq et al. (1975) suggest that Delphi is appropriate in cases where an incomplete state of knowledge about a problem or phenomenon is observed and it can be used to contribute to its enrichment; and Murphy et al. (1998) support that Delphi is appropriate where there is a lack of empirical evidence.

Considering these opinions about Delphi applicability, it has been concluded that the most appropriate methodology for this thesis is a Delphi-variant technique with the major arguments being: (a) a desire to maximize the benefit of the available pool of local SOA experts and (b) to incorporate the innovativeness of the research topic.

Also, in an attempt to reap the maximum potential outcomes from the current Delphi study, both the strengths and the limitations of the method (as presented in the previous paragraph) have been considered when designing its overall structure and the selection of the experts' panel and the codification team.

4.2.2 Delphi Study Structure

To determine the appropriate number of iterations for the proposed Delphi study both the aim of the study and the experiences from similar studies were considered. Time constraints indicated 'the sooner the better' but this had to be balanced such that a meaningful and creative input was provided to panel experts and at the same time the best possible outcomes were obtained. Erffmeyer et al. (1986), in a study into the optimal number of rounds, achieved stability after the fourth round. In more recent studies, Mulligan (2002), Powell (2003) and Richards and Curran (2002) considered three rounds were appropriate, whilst Murphy et al. (1998) and van de Ven and Delbecq (1974) suggest two or more and Loo (2002), three to four. These suggestions, combined with the aims of the study, had led to the development of a three-rounds Delphi study for the purposes of this work.

4.2.3 Codification and Experts Panel Teams Selection; Ethical Issues Addressed Through Standards, Restrictions and Rules

Setting some rules and restrictions both for the panel of experts selection and for the codification team and the way they would be functioning led to an elimination of any ethical issues that might be raised and that, in turn, to an increased reliability and accuracy for the work.

4.2.3.1 Codification Team

I was in charge of the execution and administration of this Delphi study. Moreover, I was playing the role of the codification team having the responsibility of consolidating panel members' responses and preparing the next round's questionnaire.

Having this role, some restrictions were imposed making sure of the undertaken study's integrity:

- I was not able to participate in the experts' panel at any stage;
- I was unaware of the identity of experts' panel members (except the experts' co-ordinator with whom I was interacting with and collaborating);
- I was not advised of any demographic details (except from the fact that they were all working in Cyprus) of the members of the experts' panel.

4.2.3.2 Experts Panel Selection

An important aspect of the Delphi studies is the selection of the panel of experts. Powell (2003) indicates that this selection will potentially determine the success of a Delphi study.

As far as the experts' classification is concerned, two primary aspects were considered: category (that is, whether the expert comes from Industry or Academia) and expertise in any of the factors (or domains) involved in this area (SOA).

To address the key design issue of "Academia versus Practice", a balance of academics and industry representatives was considered for the category classification. Industry practitioners were further assessed on the basis of their specialization, i.e. the aim was to include representatives from both IT consulting organizations and organizations from any field interested in adopting such the proposed model (SOA MM).

Whilst some attempts were made to classify experts based on the factors identified by the various SOA MM in existence, that is, Architecture, Infrastructure, Governance, etc., this was left to the participants to self-assess their expertise and to nominate for the factors in which they felt best qualified to participate. Self-nomination for participation would potentially increase motivation and commitment to the study, but also would not pre-establish the factors (or domains) that would be included in the proposed SOA MM (which needs to be derived from the execution of the Delphi study).

In determining the appropriate number of experts for inclusion in the panel, both the relevant literature and the aims of the study were considered. Between 15 and 20 experts was considered to be an appropriate number and was consistent with guidelines from other researchers (Loo 2002; Okoli and Pawlowski 2004; Powell 2003; Richards and Curran 2002).

Besides the above set of criteria for the experts' panel selection, an additional experts' attribute is the fact that they were all working at Cyprus. The time frame for completing this work was very limited, so the idea for considering SOA experts (both from the industry and academia) from abroad was abandoned. The "local" or "global" nature of the proposed SOA MM which was derived from local experts only, was checked, as already mentioned, by conducting interviews, as an integral part to Delphi, with some of the panel experts and local business practitioners (business owners and business practice managers).

To comply with the restrictions and rules set for the current Delphi session, the final selection of the members of the experts' panel was left to a colleague from academia who was chosen because of his understanding of qualitative research methods and deep knowledge on SOA and who was my contact point and the coordinator of the experts panel during the Delphi execution.

References

Adler, M., & Ziglio, E. (1996). *Gazing into the oracle*. Bristol, PA: Jessica Kingsley Publishers.

Ament, R. (1970). Comparison of Delphi forecasting studies in 1964 and 1969. *Futures*, March 1970.

Ascher, W., & Overholt, W. (1983). *Strategic planning and forecasting*. New York: Wiley.

Bass, B. M. (1983). Organizational decision making. In L. L. Cummins, E. Kirby Warren, & J. F. Mee (Eds.), *The Irwin series in management and the behavioral sciences*. Homewood, IL: Richard D. Irwin.

Cornish, E. (1977). *The study of the future*. Washington, D.C.: World Future Society.
Czinkota, M., & Ronkainen, I. (1997). International business and trade in the next decade: Report from a Delphi study. *Journal of International Business Studies, 28*(4), 827–844.
Delbecq, A. L., van de Ven, A. H., & Gustafson, D. H. (1975). *Group Techniques for Program Planning: A Guide to Nominal and Delphi Processes*. Glenview, III: Scott, Foresman and Company.
Erffmeyer, R. C., Erffmeyer, E. S., & Lane, I. M. (1986). The Delphi technique: An empirical evaluation of the optimal number of rounds. *Group and Organisation Studies, 11*(1–2), 120–128.
Fowles, J. (1978). *Handbook of futures research*. Connecticut: Greenwood Press.
Gatewood, R., & Gatewood, E. (1983). The use of expert data in human resource planning: guidelines from strategic forecasting. *Human Resource Planning, 5*(1), 83–94.
Gordon, T., & Hayward, H. (1968). Initial experiments with the cross-impact matrix method of forecasting. *Futures, 1*(2), 100–116.
Halal, M., Kull, B., & Leffman, A. (1997). Emerging technologies: What's ahead for 2001–2030. *The Futurist, 31*, 20–28.
Helmer, O. (1967). *Analysis of the future: The Delphi method*. Santa Monica: The RAND Corporation.
Helmer, O. (1983). *Looking forward: A guide to futures research*. Beverly Hills: Sage Publications.
Johnson, D., & King, M. (1988). *Basic forecasting techniques*. London: Butterworths.
Linstone, H., & Turoff, M. (1975). *Introduction in the Delphi method: Techniques and applications*. London: Addison-Wesley Publishing Company.
Loo, R. (2002). The Delphi method: a powerful tool for strategic management. *Policing an International Journal of Police Strategies and Management, 25*(4), 762–769.
Makridakis, S., & Wheelwright, S. (1978). *Forecasting: Methods and applications*. Santa Barbara, California: Wiley.
Martino, J. (1978). *Technological forecasting for decision making*. New York: Elsevier.
Masini, E. (1993). *Why futures studies?*. London: Grey Seal.
Mulligan, P. (2002). Specification of a capability-based IT classification framework. *Information and Management, 39*, 647–658.
Murphy, M. K., Black, N. A., Lamping, D. L., Murphy, M. K., McKee, C. M., Sanderson, C. M. B., et al. (1998). Consensus development methods, and their use in clinical guideline development. *Health Technology Assessment, 2*(3), 1–88.
Okoli, C., & Pawlowski, S. D. (2004). The Delphi method as a research tool: an example, design considerations and applications. *Information and Management, 42*, 15–29.
Ono, R., & Wedemeyer, D. (1994). Assessing the validity of the Delphi technique. *Futures, 26*(3), 289–304.
Powell, C. (2003). The Delphi technique: Myths and realities. *Journal of Advanced Nursing, 41*(4), 376–382.
Richards, J. I., & Curran, C. M. (2002). Oracles on advertising: Searching for a definition. *Journal of Advertising, 31*(2), 63–76.
Rowe, G., & Wright, G. (1999). The Delphi technique as a forecasting tool: Issues and analysis. *International Journal of Forecasting, 15*(4), 353–375.
Schmidt, R. C. (1997). Managing Delphi surveys using nonparametric statistical techniques. *Decision Science, 28*(3), 763–774.
Skulmoski, G., & Hartman, F. (2002). The Delphi method: Researching what does not exist (yet). *Proceedings of the International Research Network on Organization by Projects, IRNOP V Conference*, Renesse, The Netherlands, 2002.
Turoff, M. (1970). The design of a policy Delphi. *Technological Forecasting and Social Change, 2*, 149–171.
Van De Ven, A. H., & Delbecq, A. L. (1974). The effectiveness of nominal, Delphi, and interacting group decision making processes. *Academy of Management Journal, 17*(4), 605–621.
Wissema, J. G. (1982). Trends in technology forecasting. *R & D Management, 12*(1), 27–36.
Woudenberg, F. (1991). An evaluation of Delphi. *Technological Forecasting and Social Change, 40*, 131–150.

Chapter 5
The Proposed SOA MM

Abstract The Delphi-variant technique employed throughout this work helped develop a new inter-enterprise SOA MM (after three-round questionnaire) which is composed of five levels of maturity and seven SOA domains (enablers): Architecture, Infrastructure, Delivery, Information, Process, Organization, Governance, along with their corresponding focus areas.

Keywords Delphi-variant technique • SOA domains (enablers) • Focus areas

5.1 Delphi Technique Applied: First-Round Questionnaire and Its Findings

The application of the Delphi technique's life-cycle started by distributing an overview of the various available SOA Maturity Models (as presented in paragraph 3.3 SOA MM and Relevant Work and in Appendix I) and the CMMI maturity model (also presented in paragraph 3.3) to the panel experts (through their coordinator).

In particular, the panel experts were asked to review the available SOA MM and comment on whether they were complete (specifically in respect to whether they considered inter-enterprise elements, factors) or whether any additions might be needed. Moreover, they were asked to propose what might be recommended to include in a new SOA MM and what form this might take. Panel experts were encouraged to draw upon their experiences, and use any historical data, research, or other available resources to help in answering the posed questions.

Upon receipt of panel experts' comments (in an unstructured and informal format), the first-round questionnaire (shown in **Appendix A: Delphi Technique Applied—First-Round Questionnaire**) was formulated based on their feedback and on what this research work was trying to investigate. That is, the responses provided by the panel experts designated the major themes of the first-round questionnaire:

P. Antoniades, *SOA Maturity Model*, SpringerBriefs in Information Systems,
DOI: 10.1007/978-3-319-02453-0_5, © The Author(s) 2014

- how many levels might be included in the newly proposed SOA MM;
- what domains might be considered as SOA enablers and recommended for inclusion in the new SOA MM;
- what are the focus areas on which the identified SOA MM domains will be evaluated;

which were complemented with questions that were trying to also address other objectives of the thesis: help the participating experts' organizations position themselves in respect to SOA (current status) and anticipate their SOA maturity in 5 years time;

Once the first-round questionnaire was prepared, it was distributed to each panel member for completion though their coordinator. When all twenty panel experts responded to the questionnaire, the panel coordinator returned the questionnaire back to the codification team (myself) which was responsible for reviewing, compiling and using the responses for developing the next rounds of questionnaires.

Even though the Delphi study in progress was in its initial stages, the findings of the first-round questionnaire (shown in **Appendix B: Delphi Technique Applied—First-Round Questionnaire Findings**) started providing initial points of convergence to the various issues this research was trying to investigate.

In particular, 60 % of the respondents preferred "Five" to the question: "Considering the various available SOA Maturity Models (distributed earlier to you) and the CMMI maturity Model (also distributed earlier to you), how many levels/stages do you think might be included to a newly proposed SOA Maturity Model?"

Of the 60, 35 % chose "Five" mainly because "The HP's and Oracle's Maturity Model is the most complete" and because of "The high success rate of SOA projects by enterprises that adopted HP's and Oracle's SOA Maturity Model, according to independent studies/surveys" (as revealed by Question 4 of the First-Round Questionnaire).

The other 25 % selected "Five" because "The proposed SOA Maturity Model need not be based on any IT vendor's SOA Maturity Model(s), but instead on an independent's body/entity (if any)" and because the "CMMI was introduced by the leading IT academic institution in the US (Carnegie Mellon University) and, as such, it is recommended as the basis of all the Maturity Models of any type/kind" (as revealed by Question 5 of the First-Round Questionnaire).

With Question 6 through Question 17, an attempt was made to come up with the domains (along with the "focus areas" on which these domains would be evaluated) which are the "SOA Enablers" and it is recommended to be included in the newly proposed SOA Maturity Model (pSOAMM).

For Question 6 an assumption was made: any of the available replies that score 10 and above, would be seriously considered for inclusion in the proposed SOA Maturity Model (those with 15 and above would definitely be part of the model). This assumption was made because of the fact that the panel of IT experts was asked to choose as many of the available replies as needed.

Based on this assumption and the results of the panel, the domains: "Architecture" (18 preferences), "Business Processes" (18 preferences), "Information or Data Format" (17 preferences), "Enabling Technologies/IT Infrastructure" (16 preferences), "Governance" (15 preferences), "Business/Organization" (15 preferences) and "Delivery Methods" (15 preferences) were identified as definite to be included in the proposed SOA Maturity Model.

The "focus areas" on which the domains (tentatively identified in Question 6) would be evaluated, were revealed starting with Question 7 (through Question 17).

As far as the domain "Business/Organization" is concerned, the panel members selected the "Whether (or not) there is and the degree of IT alignment with Business strategy" (37.7 %), the "Whether (or not) there is and the degree of business involvement and understanding of SOA activities/projects" (33.3 %), and the "Change management (how easily the employees will adopt the change to SOA)" (22.2 %), as the most representing focus areas.

"The scope (inter-enterprise, enterprise-wide, department-wide, etc.) of organized SOA efforts" (33.3 %), the "Whether (or not) and to what extent there are policy enforcement schemes, reporting and exception handling guidelines/procedures across the collaborating enterprises" (33.3 %) and the "Whether (or not) there is and the degree of funding for projects" (28.9 %) were the focus areas the panel recommended be considered as the key evaluators as to whether the "Governance" domain had reached (or not) a certain level/stage of maturity.

For the "Architecture" domain, the "Whether there is an Enterprise Architecture (and which is its role) in the organization and whether it also governs the organization's relationships with its collaborating organizations" (29.6 %), the "Whether (or not) there is and the degree to which the Enterprise Architecture is being considered in the enterprise-wide and inter-enterprise planning activities" (27.7 %) and the "Whether (or not) there is and the degree to which services are cataloged and reused across the collaborating organizations" (25.9 %) were the top three choices of the panel.

The panel chose: "Whether the right tools (for software design, development, deployment, etc.), technologies are in place", "Whether industry standards are being followed (in respect to all aspects of IT operations)" and "Whether security, monitoring, and management tools (solutions) are in place," all with the same percentage of respondents' preference, 29.1 %, as focus areas for the "Enabling Technologies/IT Infrastructure" domain.

An equal percentage of respondents, 33.3 %, also preferred the focus areas for the "Delivery Methods" domain: "Whether there are any policies, practices (throughout the collaborating enterprises) for delivering the SOA solutions", "Whether there are skills and expertise (throughout the collaborating enterprises) for delivering the SOA solutions" and "Whether there are any methodologies, modeling and abstraction techniques (throughout the collaborating enterprises) for delivering the SOA solutions".

In respect to the "Information or Data Format" domain, the focus areas proposed were: "Whether the collaborating enterprises are using industry standards for the representation of data (e.g. XML)" (35.3 %), "Whether the collaborating enterprises are not only using industry standards for the representation of data (like XML), but also other canonical formats and metadata management standards (like XSD, XSLT, XPath, Expression Language, etc.)" (35.3 %), and "Whether the data is in a single location (that is, whether there is any kind of data consolidation in place throughout the collaborating enterprises)" (29.4 %).

The "Whether any kind of business process automation is in place" (37 %), the "Whether there is any kind of business process monitoring tool used (throughout the collaborating enterprises)" (27.7 %) and the "Whether a cross-application orchestration of business processes has been deployed (throughout the collaborating enterprises)" (27.7 %) were the top three options selected from those available as the focus areas for evaluating the "Business Processes" domain.

Coming into the second set of questions, where an attempt was made to establish the current and the future status of SOA (in respect to the proposed SOA Maturity Model), the replies have shown that 50 % of the panel members either "Use selectively, without a clear strategy" (30 %) or "Have an enterprise-level strategy and commitment for SOA" (20 %).

The other 50 % either "Will pursue within the next 12–24 months" (25 %) or "Not pursuing, and no immediate plans to do so" (25 %).

These results have shown that the SOA adoption in the local market is halfway to being fully utilized/used.

Also, an interesting finding in respect to SOA projects' applicability is revealed by the question "What are you currently using SOA for?" In particular, according to panel members, SOA projects are mainly encountered in Internal and External Integration activities (at 50 and 20 % respectively) but also in "Strategic business transformation" activities (30 %).

Moreover, the top three factors that facilitate the "penetration" of SOA within their organizations are: the "SOA enhances agility and flexibility; important ingredients for our organization" (33.3 %), the "SOA reduces information systems integration complexity" (33.3 %), and the "SOA lowers costs by re-using existing IT assets" (33.3 %).

It is interesting to compare these results with those derived from a similar question raised in a questionnaire distributed to the same group of participants for the purposes of a similar project undertaken eighteen months ago.

Then (eighteen months ago), to the panel members participating in that work, the top three factors that facilitated the "penetration" of SOA within their organizations were: "SOA enhances agility and flexibility; important ingredients for our organization" (50 %), "SOA reduces information systems integration complexity" (20 %), and with equal preference among respondents, at 10 %, the following: "SOA eliminates inefficiencies by lowering operating costs", "SOA creates new business opportunities through collaboration, outsourcing, and by enabling global presence for our goods and services", and "SOA accelerates growth by facilitating the integration of systems of acquired companies."

Out of these two studies, it is important to point out the fact that both the reply "SOA enhances agility and flexibility; important ingredients for our organization" and "SOA reduces information systems integration complexity" were among the top preferred replies (with some changes in the popularity between this study and the one carried out in the past) and also the fact that the third most popular reply in the current study is "SOA lowers costs by re-using existing IT assets" which was not among the top three replies in the study performed earlier.

This means that the replies "SOA enhances agility and flexibility; important ingredients for our organization" and "SOA reduces information systems integration complexity" are steadily among the most popular to the panel members.

Moreover, during the time elapsed from the previous work to this one, the panel experts appeared to realize that one of the most important benefits of SOA is re-usability; that is, the fact that they are not "directed" to rip-and-replace their existing IT infrastructure, but to re-use it.

On the contrary, the top three prohibiting factors for the adoption of SOA (and Web Services) within organizations were found to be the same as those found in the previous work (which, as said earlier, took place 18 months ago and was distributed to the same group of participants for the purposes of another similar project): "There is a lack of skills for the implementation and deployment of SOA-based Information Systems" and "lack of funding" (at 33.3 % of respondents' preference) and "the technology is still immature" and there is a "lack of anticipated ROI" (at 16.7 % of respondents' preference).

In the previous work, the prohibiting factors for the adoption of SOA (and Web Services) within organizations were found to be: "A lack of skills for the implementation and deployment of SOA-based Information Systems" (40 %), "the technology is still immature", "lack of anticipated ROI", and "lack of funding" all at 20 % of respondents' preference.

Comparing the two projects, that is this one and the one which took place 18 months ago, it is observed that the reply "There is a lack of skills for the implementation and deployment of SOA-based Information Systems" was reduced from 40 % of respondents' preference to 33.3 %, which indicates a trend for IT professionals (in general) to acquire the skills needed to minimize the gap between "demand and supply" for this new IT paradigm (SOA).

In addition, the replies: "the technology is still immature" and the "lack of anticipated ROI" were reduced to 16.7 % of respondents' preference (from 20 %) suggesting that the technology (SOA/Web Services) is becoming more and more mature and that the first reliable testimonials on "returns-on-SOA-related-investments" have started persuading people about this technology's business value.

What is also important to point out is the fact that the reply "lack of funding" rose to 33.3 % of respondents' preference in the current work, from 20 % in the previous one. This, of course, indicates that funding of SOA projects is still a problem (that is, cannot be easily secured), but at the same time this also means that -most probably- many more requests were made for SOA activities/projects (since the previous work) and they were rejected. This is a reasonable explanation to the reply's "lack of funding" increasing trend.

As with the previous work, a different question was raised concerning the SOA "penetration" 5 years from now in the broader business landscape to the one concerning the SOA "penetration" in 5 years within the panel experts' organizations.

Of interest, though, in both studies is the fact that the panel members responded in the exact same way in the current study as in the previous one: they believe that the status of SOA 5 years from now within their enterprises and in the broader business landscape will be the same. In both questions, the percentage of respondents' preference replying: "there will be a considerable (to great) increase in the adoption of SOA (and Web Services)" in this study was exactly the same as the previous one (50 %). What has changed, though, is the percentage of respondents' preference to the replies: "SOA (and Web Services) will be the dominant technology in 5 years time, like what the 'static' Web is today" (35 % now, from 25 %) and "there will be a slight (or no) increase in the adoption of the technology" (15 % now, from 25 %) which clearly shows the increased "trust" of people in SOA.

Both in the previous study and in the current one, the top facilitators of the SOA adoption 5 years from now were considered: the "greater business flexibility and agility", the "lowering of costs of integrating existing applications and systems", and the fact that "there will be proven success stories associated with the application of SOA-based systems in highly demanding environments indicating the inherent benefits of the technology." The surprising feature, though, in this question, is the fact that the percentages of respondents' preferred replies in the current study were almost the same as those provided in the previous one (33.3 now vs. 33.3 then, 33.3 now vs. 33.3 then, 20 now vs. 19.6 then).

So, 1.5 years from the time the previous work conducted, the panel members continue to believe that the SOA adoption facilitators are the same as the ones chosen in the previous study.

The prohibiting factors for the adoption of SOA-based systems, 5 years from now, as revealed by the current study, are considered to be: "the skills needed to implement and use the technology will be very limited in respect to the capacity required by the technology" (44.4 %) and "the ROI will never be encouraging enough" along with "the technology will fail to prove itself when it will be implemented and deployed in systems where high security features are mandated" both at 22.2 % of preferences.

The previous study, though, was slightly different from the current one on this question. According to the previous study, the prohibiting factors for the adoption of SOA-based systems, in 5 years time, were considered to be: "the skills needed to implement and use the technology will be very limited in respect to the capacity required by the technology" (40 %), "the ROI will never be encouraging enough", "the technology will fail to prove itself when it is implemented and deployed in systems where high security features are mandated", and "the big supporters of the technology—Big IT companies like Microsoft, IBM, Oracle, etc.—will abandon the technology and they will look for more profitable technologies/solutions", all three with 20 % of preferences.

What is important to stress, if a comparison between the two studies was done in respect to this question, is the fact that the IT people (who were the panel experts) started to believe that the big IT vendors "are here to stay", as far as SOA technology, is concerned. That's the reason why the reply "The big supporters of the technology (Big IT companies like Microsoft, IBM, Oracle, etc.) will abandon the technology and they will look for more profitable technologies/solutions" decreased to 11.1 % of respondents' preference in the current study (in respect to 20 % of the previous one).

The findings of the first-round questionnaire were also distributed to the panel experts who were allowed to review their responses in light of the opinions of other experts, add comments, and change their responses if desired. The codification team then reviewed the responses and used this information to develop more specific questions to be used in the second questionnaire (shown in **Appendix C: "Delphi Technique Applied—Second-Round Questionnaire"**).

5.2 Delphi Technique Applied: Second-Round Questionnaire and Its Findings

The new questions (which constituted the second-round questionnaire) formulated by the codification team were posed to the panel (through the experts panel coordinator). Upon completion, panel members returned their answers back to the codification team. The findings of the second-round questionnaire started showing a consensus among panel experts about the topics the research was trying to investigate.

Five levels/stages might be included in the new SOA Maturity Model mainly because "the HP's and Oracle's Maturity Model is the most complete" and because of "the high success rate of SOA projects by enterprises that adopted HP's and Oracle's SOA Maturity Model, according to independent studies/surveys."

Moreover, the choice of five levels/stages in the proposed SOA Maturity Model is further supported by the arguments: "The proposed SOA Maturity Model need not be based on any IT vendor's SOA Maturity Model(s), but instead on an independent's body/entity model (if any)" and because the "CMMI was introduced by the leading IT academic institution in the US (Carnegie Mellon University) and as such it is recommended as the basis of all the Maturity Models of any type/kind."

Another important piece of information "captured" by the panel experts is the interpretation that could be given to these five levels of the proposed SOA Maturity Model (pSOAMM).

They agreed that their choice to select five as the ideal number of levels to be included in the pSOAMM also implied that not only the number of levels of the proposed SOA Maturity Model might be the same as those of HP's and Oracle's and CMMI's, but also the interpretation provided by these models for each of the levels would resemble that of the pSOAMM.

This means that when, for instance, an enterprise is in Level 1, SOA is a relatively new concept to it. Level 2 means that the enterprise has committed to adopting SOA for certain parts of the organization. SOA Maturity Level 3 is "awarded" to enterprises that have adopted SOA as a strategic enterprise-wide architectural paradigm. Level 4 means that SOA is fundamental to the way the enterprise operates and services might be used outside the enterprise (for instance, to be "consumed" by business partners or customers). When an enterprise reaches Level 5 maturity, it means that this enterprise operates a dynamic SOA with business and IT synchronized to achieve an optimum balance of agility, performance, risk and cost.

To be able to classify an enterprise at a certain SOA level, though, a set of domains are required. In essence, domains are capability areas where an enterprise must focus to increase SOA maturity. Each domain requires a set of capabilities at each maturity level. Within a domain, capabilities are centered around "focus areas" on which the domains are evaluated.

The panel members recommended that the domains (or SOA enablers) be included in the proposed SOA Maturity Model (pSOAMM) are: "Architecture", "Business Processes", "Information or Data Format", "Enabling Technologies/IT Infrastructure", "Governance", "Business/Organization" and "Delivery Methods".

As far as the "Business/Organization" domain is concerned, the panel members selected the "Whether (or not) there is and the degree of IT alignment with Business strategy", the "Whether (or not) there is and the degree of business involvement and understanding of SOA activities/projects", and the "Change management (how easily the employees will adopt the change to SOA)" as the most representing focus areas for this domain.

"The scope (inter- enterprise, enterprise-wide, department-wide, etc.) of organized SOA efforts", the "Whether (or not) and to what extent there are policy enforcement schemes, reporting and exception handling guidelines/procedures across the collaborating enterprises" and the "Whether (or not) there is and the degree of funding for projects" were the focus areas the panel members recommended be considered as the key evaluators as to whether the "Governance" domain has reached (or not) a certain level of maturity.

For the "Architecture" domain, the "Whether there is an Enterprise Architecture (and which is its role) in the organization and whether it also governs the organization's relationships with its collaborating organizations", the "Whether (or not) there is and the degree to which the Enterprise Architecture is being considered in the enterprise-wide and inter-enterprise planning activities" and the "Whether (or not) there is and the degree to which services are cataloged and reused across the collaborating organizations" were the top three choices of the panel regarding the focus areas of this domain.

The panel chose: "Whether the right tools (for software design, development, deployment, etc.) and technologies are in place", "Whether industry standards are being followed (in respect to all aspects of IT operations)" and "Whether security, monitoring, and management tools (solutions) are in place" as focus areas for the "Enabling Technologies/IT Infrastructure" domain.

The recommended focus areas for the "Delivery Methods" domain were: "Whether there are any policies and practices (throughout the collaborating enterprises) for delivering the SOA solutions", "Whether there are skills and expertise (throughout the collaborating enterprises) for delivering the SOA solutions" and "Whether there are any methodologies, modeling and abstraction techniques (throughout the collaborating enterprises) for delivering the SOA solutions," according to panel preference.

In respect to the "Information or Data Format" domain, the focus areas proposed were: "Whether the collaborating enterprises are using industry standards for the representation of data (e.g. XML)", "Whether the collaborating enterprises are not only using industry standards for the representation of data (like XML), but also other canonical formats and metadata management standards (like XSD, XSLT, XPath, Expression Language, etc.)" and "Whether the data is located on a single location (that is, whether there is any kind of data consolidation in place throughout the collaborating enterprises)."

The "Whether any kind of business process automation is in place", the "Whether there is any kind of business process monitoring tool used (throughout the collaborating enterprises)" and the "Whether a cross-application orchestration of business processes has been deployed (throughout the collaborating enterprises)" were the top three options selected from those available as the focus areas for evaluating the "Business Processes" domain.

Capturing the maturity levels and the SOA domains (along with their focus areas) from the panel members who participated in this study (as revealed by the findings of the "Delphi Technique Applied—Second-Round Questionnaire") led to the "birth" of the proposed SOA Maturity Model (pSOAMM) as shown in Table 5.1 below.

The proposed model captures the need identified for an inter-enterprise SOA MM by incorporating the appropriate focus areas for each of the domains in the model. In particular, it is the focus areas (and the way these were evolved through the Delphi questionnaires) along with their corresponding benchmarking scheme adopted and included in the model (Table 5.1) *that considers all those elements that measure the capability of how well an organization is interacting with its collaborating organizations' resources, in respect to SOA.*

It is true that the domains included in the proposed SOA Maturity Model ("Architecture", "Business Processes", "Information or Data Format", "Enabling Technologies/IT Infrastructure", "Governance", "Business/Organization" and "Delivery Methods") may also be encountered in other SOA MM available, but none of them includes focus areas and a corresponding benchmarking scheme which captures and measures the capability of the organization to capitalize on connecting and aligning its operations with other (collaborating) companies to meet important organizational goals using one the latest IT evolutions: Service-Oriented Architecture, as the proposed SOA MM does (Table 5.1).

For instance, if we were to determine the maturity level of the "Governance" domain in an organization (both internally and in respect to its interaction with

Table 5.1 The pSOAMM including the maturity levels and SOA domains (SOA enablers) along with their focus areas

	Maturity Levels				
	Level 1	Level 2	Level 3	Level 4	Level 5
pSOAMM's SOA enabler /domain (along with its focus areas)					
Architecture (*Focus areas:* a. Role of enterprise architecture;	a. No formal enterprise architecture;	a. There is an architecture teams within the collaborating organizations, but their impact on their enterprises is not well understood;	a. Heavy investments from the collaborating organizations in enterprise architecture, but they are only at the beginning of realizing associated benefits;	a. Enterprise architecture proves its value through business results;	a. Enterprise architecture transforms IT into business value;
b. Cataloging and reuse;	b. No cataloging or reusing web services;	b. No formal catalog of web services, but reuse is taking place informally;	b. There is a formal method for registering and finding web services;	b. The use of web services repositories is enforced;	b. Advanced users of web services repositories;
c. Planning and guidance)	c. Enterprise architects are not involved in planning;	c. Architects offer their advice, but their influence in planning is minimal;	c. Enterprise architects have a fair degree of authority;	c. Enterprise architects are involved in every line of business within the collaborating organizations;	c. Enterprise architects have a seat at the board room table within the collaborating organizations;

(continued)

Table 5.1 continued

pSOAMM's SOA enabler /domain (along with its focus areas)	Maturity Levels Level 1	Level 2	Level 3	Level 4	Level 5
Infrastructure (*Focus areas:* a. standards;	a. No any formal stance regarding standards;	a. IT assets started to be rationalized into standards-based versus non-standards;	a. A formal position is held on which standards are important;	a. Adherence to standards plays a critical role for all projects, particularly those spread throughout the collaborating organizations;	a. Continuous investment in emerging standards;
b. Security and monitoring;	b. Individual systems have their own security and monitoring models; little has been done to unify them;	b. Security is addressed on a case-by-case basis;	b. Started standardizing the security schemes of the various systems across the collaborating organizations;	b. Centralized security mechanisms have been fully deployed;	b. Security has been completely abstracted out of all systems; compliance to all security regulatory standards;
c. Management and operations	c. Individual systems have their own separate management and operations mechanisms; little has been done to consolidate them;	c. Automation techniques started to be utilized to make systems more manageable;	c. Common management platform for all systems in the collaborating organizations is in place, but coverage is not complete;	c. Management platform has been fully deployed;	c. Management has been consolidated into a unified console, allowing easy troubleshooting and monitoring;

(continued)

Table 5.1 continued

pSOAMM's SOA enabler /domain (along with its focus areas)	Maturity Levels Level 1	Level 2	Level 3	Level 4	Level 5
Delivery					
(*Focus areas:* a. Project versus inter-enterprise focus;	a. Each new application is treated as an independent new project;	a. New applications are subject to review and approval by an architectural review board which is composed of members from the collaborating organizations;	a. Project changes may be mandated in the interest of maximizing benefits of the collaborating organizations (inter-enterprise focus);	a. The disciplined inter-enterprise focus has resulted in significant service reuse, faster time-to-market, and less custom development;	a. New application development is minimal, being replaced by rapid "orchestration" of services located in the inter-enterprise repository;
b. Skills and methodologies in place;	b. No formal assessment has been performed for SOA skills and methodologies;	b. SOA-related skill-set assessment is currently taking place;	b. A baseline for skills and methodologies has been established and it is standardized across the collaborating enterprises;	b. The required skill levels and methodologies are in place throughout the collaborating organizations;	b. Continuous investment in new skills and methods;
c. Modeling and abstraction techniques)	c. Limited and inconsistent modeling and abstraction techniques across projects;	c. Modeling and abstraction techniques are under development;	c. Modeling and abstraction techniques are employed in all new projects;	c. Modeling and abstraction techniques expanded to be used in legacy systems, too;	c. Modeling and abstraction techniques are inextricable parts of every system throughout the collaborating organizations;

(continued)

Table 5.1 continued

	Maturity Levels				
pSOAMM's SOA enabler /domain (along with its focus areas)	Level 1	Level 2	Level 3	Level 4	Level 5
Information					
(*Focus areas:* a. Data standards and canonical formats;	a. No data standards or canonical data formats are in place;	a. Design has begun on data representation standards and canonical formats;	a. Inter-enterprise data model is under construction;	a. Plan for data standardization is complete and fully deployed;	a. Plan continuously evolves to address new industry requirements;
b. Metadata management;	b. Metadata management is not viewed as important;	b. Design has begun on a metadata management solution;	b. Plan for a metadata management solution is implemented for specific types of business data;	b. Plan for metadata management is complete and fully deployed;	b. Data management plan is adjusted as new requirements evolve;
c. Single source of truth)	c. No common data model in place for any application;	c. A common data model has been constructed but is in limited use;	c. Multiple applications are using a common data model;	c. All applications are tied into the common data model, but data duplication and cleansing are still an issue;	c. All applications are tied into the common data model; no data duplication is encountered or data cleansing is needed;

(continued)

Table 5.1 continued

	Maturity Levels				
	Level 1	Level 2	Level 3	Level 4	Level 5
pSOAMM's SOA enabler /domain (along with its focus areas)					
Process (*Focus areas*: a. Process automation;	a. Process automation is taking place as an ad-hoc response, not a strategic plan;	a. Some process modeling techniques are being used;	a. Process automation techniques are being used in the majority of lines of business (throughout the collaborating organizations);	a. Process automation techniques have become standardized across most lines of business (throughout the collaborating organizations);	a. Automation capabilities have advanced to an optimum degree throughout the collaborating organizations;
b. Composite application development;	b. Limited composite application development;	b. Growing capabilities for composite application development is encountered;	b. Repeatable success across some lines of business (throughout the collaborating organizations);	b. Repeatable success across all lines of business (throughout the collaborating organizations);	b. Application development has transformed into application assembly, leading to rapid time to market and ideal levels of flexibility;
c. Process measurement and scoring)	c. Process measurement and scoring are loosely defined;	c. Process measurement and scoring metrics have started to be in place throughout the collaborating organizations;	c. Meaningful metrics have been established, but only for a limited set of processes;	c. Meaningful metrics have been established across all key business processes;	c. Metrics have been established across all business processes;

(continued)

Table 5.1 continued

	Maturity Levels				
	Level 1	Level 2	Level 3	Level 4	Level 5
pSOAMM's SOA enabler /domain (along with its focus areas)					
Organization (*Focus areas:* a. IT alignment with business strategy;	a. IT and business groups across the collaborating organizations do not have frequent contacts;	a. IT and business started discussing the impact of web services on the business;	a. IT and business closely collaborate on supporting key business processes;	a. IT understands business processes extremely well and it is responsive to most new business requirements;	a. IT is perceived as very critical for the planning and execution of critical business strategy, innovation and differentiation in the market;
b. Change management;	b. SOA impact in the collaborating organizations is not well understood;	b. Started learning how SOA will change the way things are carried out;	b. Partial understanding of SOA's impact in the collaborating organizations;	b. Understanding of SOA's value to business and how this is affecting the majority of the business processes;	b. Clear understanding of SOA's value to business which leads to a change management strategy for driving SOA throughout the collaborating enterprises;
c. Business involvement and understanding)	c. Only IT departments or people (of the collaborating organizations) are involved with the maintenance of web services;	c. Business people are beginning to understand web services potential impact;	c. Gain greater understanding of the business and IT benefits expected to accrue from reusable services;	c. Tangible and intangible benefits derived by shared services used in the major lines of business are realized by business people;	c. Clear understanding by business people of the costs and benefits of SOA across all levels of the collaborating enterprises;

(continued)

Table 5.1 continued

pSOAMM's SOA enabler /domain (along with its focus areas)	Maturity Levels Level 1	Level 2	Level 3	Level 4	Level 5
Governance (*Focus areas:* a. Funding and accounting;	a. Each business unit in the collaborating organizations treats web services as a standalone cost because there is no service reuse;	a. Limited service reuse across the collaborating enterprises; no formal cost allocation model has been formulated;	a. Service reuse is occurring across the collaborating enterprises, but no formal cost allocation model is used;	a. Service reuse is occurring across the collaborating enterprises and there is a formal cost allocation model in place;	a. Careful metering of shared service usage allows effective models of budget distribution across all business units of the collaborating enterprises;
b. Scope of organized SOA efforts;	b. Only certain people in the IT department(s) of the collaborating organizations are aware of SOA;	b. Service awareness is pervasive throughout the IT departments of the collaborating organizations;	b. Service awareness spreading to business process owners in certain lines of business of the collaborating organizations;	b. Services strategy is in place and followed by most of the lines of business of the collaborating organizations;	b. The services strategy is understood and mandated across all lines of business of the collaborating organizations;
c. Policies, reporting, and exception handling)	c. They are treated as additional work for IT and are handled on an ad-hoc basis;	c. Roles and responsibilities have been formally modeled to address policy management and reporting;	c. Individual business units (in the collaborating organizations) have centralized their policy enforcement, reporting, and exception handling;	c. A single centralized team has been formed and it is responsible over policy, compliance, and exception handling;	c. A central policy team can dynamically implement rule and policy changes independent of underlying IT systems;

collaborating enterprises), the key evaluators (focus areas) on which our "judgment" would be based on were identified to be:

- "Whether (or not) there is and the degree of funding for projects";
- "The scope (inter-enterprise, enterprise-wide, department-wide, etc.) of organized SOA efforts";
- "Whether (or not) and to what extent there are policy enforcement schemes, reporting and exception handling guidelines/procedures across the collaborating enterprises";

In particular, this means that when an enterprise is in Level 1 (that is, SOA is a relatively new concept in respect to "Governance" domain) the following capability level holds for each of the focus areas respectively:

- "Each business unit in the collaborating organizations treats Web Services as a standalone cost because there is no service reuse";
- "Only certain people in the IT department(s) of the collaborating organizations are aware of SOA";
- "They are treated as additional work for IT and are handled on an ad-hoc basis";

To the other end, Level 5 means that the enterprise operates a dynamic SOA with business and IT synchronized to achieve an optimum balance of agility, performance, risk and cost, in respect to "Governance" domain. In particular, the capability level for each of the corresponding focus areas is as follows:

- "Careful metering of shared service usage allows effective models of budget distribution across all business units of the collaborating enterprises";
- "The services strategy is understood and mandated across all lines of business of the collaborating organizations";
- "A central policy team can dynamically implement rule and policy changes independent of underlying IT systems";

Levels 2–4 (in respect to "Governance" domain) fall within Levels 1 and 5 extremes and relevant capability levels for the focus areas hold and are as shown in Table 5.1.

The remaining of the findings of the "Delphi Technique Applied—Second-Round Questionnaire" helped conclude on the other research objectives.

Trying to establish the current status of SOA within the panel experts' enterprises (but also of the broader business landscape), the replies provided by the panel members have shown that the majority either "use selectively, without a clear strategy," or "have an enterprise-level strategy and commitment to SOA," or "will pursue within the next 12–24 months".

Only a quarter of the respondents replied "Not pursuing, and no immediate plans to do so" in respect to their "relationship" to SOA.

As far as the future status (in particular, after 5 years) of SOA is concerned, a very large number of respondents replied that "there will be a considerable (to great) increase in the adoption of SOA (and Web Services)" and that "SOA (and Web Services) will be the dominant technology 5 years from now, like what the –static- Web is today."

Some other interesting findings were also revealed by the second-round questionnaire in respect to SOA. For instance, to the panel members, SOA projects are mainly observed in "Internal and External Integration" activities but also in "Strategic business transformation" activities.

Moreover, the top three factors that facilitate the "penetration" of SOA within panel members' organizations are: the "SOA enhances agility and flexibility; important ingredients for our organization," the "SOA reduces information systems integration complexity," and the "SOA lowers costs by re-using existing IT assets."

Even 5 years from now, the top facilitators in the SOA adoption/penetration were considered to be the "greater business flexibility and agility", the "lowering of costs of integrating existing applications and systems", and the fact that "there will be proven success stories associated with the application of SOA-based systems in highly demanding environments indicating the inherent benefits of the technology."

On the contrary, the top three prohibiting factors for the adoption of SOA (and Web Services) -today- within organizations were found to be: "There is a lack of skills for the implementation and deployment of SOA-based Information Systems", the "lack of funding", "the technology is still immature" and the "lack of anticipated ROI."

The prohibiting factors for the adoption of SOA-based systems, even 5 years from now, as revealed by the second-round questionnaire, are considered to be: "the skills needed to implement and use the technology will be very limited in respect to the capacity required by the technology", "the ROI will never be encouraging enough" and "the technology will fail to prove itself when it is implemented and deployed in systems where high security features are mandated."

Due to the fact that the objective: "what the status of SOA will be within enterprises in the future, in respect to the proposed Maturity Model" was not readily derived from this second-round questionnaire, a third-round was required and executed.

5.3 Delphi Technique Applied: Third-Round Questionnaire and Its Findings (Application/Use of the Proposed Inter-Enterprise SOA MM)

In this round, the panel members were asked to designate the maturity level of their enterprises on each of the SOA domains identified in the derived SOA Maturity Model (which was the outcome of the second-round questionnaire) both for the current status of SOA and the status of SOA in 5 years.

The proposed SOA Maturity Model (shown on the previous page as Table 5.1) was handed-out to the panel members along with the third-round questionnaire (shown as **Appendix D: Delphi Technique Applied—Third-Round Questionnaire**)

In particular, the panel members were called to designate their choices by drawing the appropriate boxes in the model indicating the maturity level of each of the available domains.

Table 5.2 Current status of SOA using the pSOAMM

	Maturity levels				
	Level 1	Level 2	Level 3	Level 4	Level 5
pSOAMM's SOA enabler/domain (along with its focus areas)					
Architecture (*Focus areas:* a. Role of enterprise architecture; b. Cataloging and reuse; c. Planning and guidance)	3/20	9/20	3/20		
Infrastructure (*Focus areas:* a. Standards; b. Security and monitoring; c. Management and operations)	5/20	6/20	4/20		
Delivery (*Focus areas:* a. Project versus enterprise focus; b. Skills and methodologies in place; c. Modeling and abstraction techniques)	8/20	4/20	3/20		
Information (*Focus areas:* a. Data standards and canonical formats; b. Metadata management; c. Single source of truth)	7/20	5/20	3/20		
Process (*Focus areas*: a. Process automation; b. Composite application development; c. Process measurement and scoring)	9/20	4/20	2/20		
Organization (*Focus Areas:* a. IT alignment with business strategy; b. Change management; c. Business involvement and understanding)	7/20	6/20	2/30		
Governance (*Focus areas:* a. Funding and accounting; b. Inter-enterprise/cross-organizational involvement; c. Policies, reporting, and exception handling)	9/20	4/20	2/20		

Table 5.3 Status of SOA in five(5) years using the pSOAMM

	Maturity levels				
	Level 1	Level 2	Level 3	Level 4	Level 5
pSOAMM's SOA enabler/domain (along with its focus areas)					
Architecture (*Focus areas:* a. Role of enterprise architecture; b. Cataloging and reuse; c. Planning and guidance)	4/20	7/20	5/20	2/20	2/20
Infrastructure (*Focus areas:* a. Standards; b. Security and monitoring; c. Management and operations)	5/20	5/20	4/20	4/20	2/20
Delivery (*Focus areas:* a. Project versus enterprise focus; b. Skills and methodologies in place; c. Modeling and abstraction techniques)	6/20	4/20	5/20	3/20	2/20
Information (*Focus areas:* a. Data standards and canonical formats; b. Metadata management; c. Single source of truth)	7/20	5/20	3/20	3/20	2/20
Process (*Focus areas*: a. Process automation; b. Composite application development; c. Process measurement and scoring)	9/20	4/20	3/20	2/20	2/20
Organization (*Focus Areas:* a. IT alignment with business strategy; b. Change management; c. Business involvement and understanding)	7/20	6/20	4/30	2/20	1/20
Governance (*Focus areas:* a. Funding and accounting; b. Inter-enterprise/cross-organizational involvement; c. Policies, reporting, and exception handling)	9/20	5/20	5/20	1/20	

Sample responses both for the current status of SOA and the status of SOA in 5 years within the panel members' enterprises are shown in **Appendix E: Delphi Technique Applied—Third-Round Questionnaire Sample Responses (as "E1. Current Status Of SOA" and "E2. Status Of SOA In 5 Years" respectively).**

Collectively, the results obtained, in respect to the current status of SOA within the panel members' enterprises (and subsequently of the majority of the local enterprises) are illustrated in Table 5.2

First of all, Table 5.2 shows that the total number of actual entries in the various maturity levels per SOA domain (provided by the panel members) is fifteen out of twenty. This justifies the 25 % of the panel members who admitted that they are "not pursuing any SOA activities, or have no immediate plans to do so."

What is also interesting to point out is the fact that the responses provided by the panel experts (concerning the SOA status-es within their organizations) are distributed between the first three levels of maturity for all the SOA domains.

This means that currently SOA is a relatively new concept to most of the panel members' enterprises, but also that some of these enterprises have committed to adopting SOA even for certain parts of the organization. A smaller number of panel members' enterprises have adopted SOA as a strategic enterprise-wide architectural paradigm.

Similarly, the results obtained in respect to the status of SOA in 5 years within the panel experts' enterprises (and subsequently of the majority of the local enterprises) are shown in Table 5.3.

Comparing the information presented in Table 5.2 to the information in Table 5.3, it can be concluded that **all** the panel experts believe that their enterprises will be "somewhere" in the roadmap toward SOA adoption, 5 years from today, as opposed to the previous finding concerning the current SOA status, where 15 out of 20 of the panel experts indicated that their enterprises are using SOA somehow.

Furthermore, it is obvious, if we were to compare the two tables (Tables 5.2 and 5.3) that the SOA penetration is expected to be greater 5 years from now. Particularly, the panel experts' responses are spread throughout all five levels of the newly proposed SOA Maturity Model. That is, it is consciously believed that the SOA will be fundamental to the way the enterprises will operate and services might be used outside the enterprise, too (for instance, to be "consumed" by their business partners or customers).

Moreover, there were responses that indicated that the panel experts expected that their enterprises would operate a dynamic SOA with business and IT synchronized in order to achieve an optimum balance of agility, performance, risk and cost, in 5 years time.

To conclude, it can be said that even though some of the panel experts' enterprises (meaning also a representative sample of the local market) are not currently using SOA, they are confident enough that they will do so within the next 5 years.

Chapter 6
Proposed SOA MM Broadness and Applicability (in Particular: "Local" or "Global" SOA MM)

Abstract In an attempt to explore the broadness and applicability of the proposed SOA MM in other regions and markets, the following conclusion could be drawn: even though the Delphi-variant technique proved to be beneficial when there is a need to obtain views and judgments from geographically dispersed knowledgeable people, the fact that the panel of experts selected for this research originated from local -only- Industry and Academia, did not affect its success and reliability because all the other factors (besides locality) were preserved: diversity of knowledge and expertise, acceptable group size whose members never functioned in a simultaneous face-to-face meeting, etc. So, this newly developed SOA MM provides a good basis and template for any other similar attempts that might be undertaken in other regions and markets. Thus, it is advised to be evaluated, reviewed and adjusted accordingly by academia and industry representatives elsewhere as well.

Keywords SOA MM • Delphi-variant technique • Broadness • Applicability • Local • Global

The final objective of this research work: to come up with a conclusion about the "local" or "global" nature of the proposed SOA MM, was fulfilled by conducting interviews with two different groups of interviewees. Firstly, the "locality" of the proposed SOA MM was checked by taking a sample population of those who contributed to the formation of the model (panel experts) and secondly, by conducting a set of interviews with selected business owners and business practice managers.

The set of points discussed is slightly different from one group to the other. In particular, the points discussed with the selected panel experts is shown as **Appendix F** (titled **"Selected Panel Experts Interview -On Proposed SOA MM Locality-"**) and in **Appendix H** (titled **"Business Practitioners -Business Owners And Business Practice Managers- Interview -On Proposed SOA MM Locality-"**) the set of points discussed with business owners and business practice managers is shown.

P. Antoniades, *SOA Maturity Model*, SpringerBriefs in Information Systems,
DOI: 10.1007/978-3-319-02453-0_6,

A good sample (3 out of 20) of the population of the panel experts was asked (again, through the expert panel coordinator who had also chosen them) to also contribute towards the final objective of this research work; that is, to determine whether the proposed SOA MM can be used by (and be beneficial to) Cypriot organizations only, or whether it is a "global" one (that is, it can also be used outside Cyprus, too).

The selection of the three experts was left to the expert panel coordinator's discretion mainly because research work's reliability and minimization of any criticisms of biases continued to be priorities to be secured throughout the whole "life" of this work.

The issues discussed with the selected experts can be found in **Appendix F** (titled **"Selected Panel Experts Interview -On Proposed SOA MM Locality-"**) and the objective was to come up with conclusions mainly on:

- whether they think that the Delphi technique is the most appropriate tool for carrying out this type of research work (proposing a new SOA MM);
- whether they think that the proposed SOA MM is appropriate for local organizations only or whether it can be used (and have a real value) by organizations outside Cyprus, too;
- the factors that make them certain that the proposed SOA MM can only be used in Cyprus, if they had chosen to reply so in the previous point;
- the ingredients that make the proposed SOA MM "global" (that is, that it can be used by organizations outside Cyprus, too), if they had chosen to reply so in one of the previous points;
- the domains that could be added to the proposed SOA MM that would make it "global" (that is, that it can be used by organizations outside Cyprus, too), if it is currently a "local" one;
- whether anything else could be added to the proposed SOA MM to make it even more complete, broad and "global";

The second group was composed of five business owners and business practice managers, one from each of the major industries contributing to the Cyprus economy (Tourism, Banking) and industries heavily invested in IT (Government, Telcos, Utilities). The identification of the organizations and subsequently the individuals from those organizations (who were asked to participate in this study), was based on the organization size -and subsequently, the organization's impact on the country's economy-, the distributiveness of its business processes (because that's where Integration/SOA-based solutions are desired) and the individual's relevance to the issue under study (SOA).

The issues discussed with the business owners and business practice managers can be found in **Appendix H** (titled **"Business Practitioners -Business Owners And Business Practice Managers- Interview -On Proposed SOA MM Locality-"**). Again, the objective was to determine whether the proposed SOA MM can be used by (and be beneficial to) Cypriot organizations only, or whether it is a "global" one (that is, it can also be used outside Cyprus, too), raising not only those points discussed with the chosen panel experts (above), but also the following:

- whether the proposed SOA MM addresses real, day-to-day business challenges, or whether they consider it to be just a theoretical framework with no (or minimal) business value;

- whether the proposed SOA MM addresses the challenge of incorporating those domains and focus areas that refer to inter-enterprise settings;

Both groups were provided a description of the proposed SOA MM, an introduction on SOA (titled: **"The Business Imperative For SOA"** -shown as **Appendix G**-), and an introduction on Delphi technique (as shown in Sect. 4.1), which helped them to address the examined points.

The findings from the interviews (both with the panel experts and the business owners and practice managers) were interesting and shed some light on the examined points. In particular, the following conclusions were drawn:

- the proposed SOA MM addresses real, day-to-day business challenges mainly because of the fact that it was formed also by the feedback provided by industry representatives (not only academia representatives) who were members of the panel of experts and as such it is likely to be of value to organizations that will use it;
- the Delphi technique seems to be the most appropriate tool for proposing a new SOA MM mainly because Delphi's strength lies in the fact that it is ideal when the research issue is complex enough, as Okoli and Pawlowski (2004) and Ono and Wedemeyer (1994) suggested, (and proposing a new SOA MM was indeed a challenging task), but also due to the fact that the existing SOA MMs lacked the "inter-enterprise"-related domains and the associated focus areas and the fact that Delphi is -also- ideal when "there is a need for further contribution to an incomplete state of knowledge around a research topic" (Delbecq et al. 1975);
- the proposed SOA MM incorporates the appropriate domains and corresponding focus areas that address the requirements observed in inter-enterprise settings;
- even though the technology itself (SOA) is universal and some of the panel experts are working for global companies with operations in Cyprus, too, and most of them studied at foreign universities (so one may infer that the proposed SOA MM could be considered "global"), the fact that there was no participant from any other country (in particular, doing business in any other area of the world) "narrows" the proposed SOA MM "broadness" and that's why it is characterized as a "local" one, with no guarantee of its direct applicability in organizations outside Cyprus, but with the potential and completeness to be beneficial to other countries' organizations too, if appropriate (hopefully minor) adjustments are implemented;
- IT services, SOA, SOA MM (and particularly the proposed SOA MM) seem to be ideal for Cyprus and any other country which has similar business infrastructure and landscape (service-oriented), mainly because:
  - in Cyprus, the business landscape is primarily service-oriented; so any IT technologies, architectures, etc., that have the same "DNA" as the business landscape (that is, the ability to constantly change, to be agile and flexible, to respond quickly, etc.), are very appealing to local organizations;
  - of the fact that the majority of Cypriot organizations are small-to-medium sized enterprises (even, family-owned enterprises) which makes them depend on many other collaborating organizations in order for a transaction or a business process to be successfully completed; that's where our proposed SOA MM could be called in to address this inter-enterprise requirement (where none of the others available can do);

- the proposed SOA MM (and the others, too) will give the opportunity to organizations, if they desire to be competitive players in the global business arena, to comply with industry standards, policies and rules by enforcing them within their organizations in order to be able to collaborate and cooperate with other global partners and suppliers;
- of the fact that the political and business landscape is in constant flux and it is very frequent the case that there is a need for organizations to comply to the European Union and other (legal, etc.) regulations (SOA along with its corresponding MMs addresses that requirement).

References

Delbecq, A. L., van de Ven, A. H., & Gustafson, D. H. (1975). *Group techniques for program planning: A guide to nominal and delphi processes*. Glenview, III: Scott, Foresman and Company.

Okoli, C., & Pawlowski, S. D. (2004). The Delphi method as a research tool: An example, design considerations and applications. *Information & Management, 42*, 15–29.

Ono, R., & Wedemeyer, D. (1994). Assessing the validity of the Delphi technique. *Futures, 26*(3), 289–304.

Chapter 7
Conclusions, Limitations and Future Research

Abstract In the proposed SOA maturity model (pSOAMM) developed, five levels of SOA maturity were identified and a set of SOA domains along with their corresponding focus areas on which the domains would be evaluated, were defined. In general, the pSOAMM can be used by enterprises to facilitate their SOA adoption because pSOAMM provides an insight into the current SOA maturity level and a roadmap on how higher levels of SOA maturity can be achieved (and thus, benefit more from further SOA initiatives). Moreover, the newly developed SOA maturity model (pSOAMM) was used to capture the current SOA status and the SOA status in 5 years within the representative sample of local enterprises (through panel experts). However, this newly proposed SOA MM needs to be adjusted accordingly by academia and industry representatives in other regions and markets to prove its broadness and globality.

Keywords pSOAMM • Five levels of maturity • SOA domains (enablers) • Focus areas • Broadness • Globaity

The major objective of this research work was to come up with a new Delphi-derived inter-enterprise SOA Maturity Model and then use this model to "capture" both the current SOA status and the SOA status in 5 years within the enterprises of the participating panel experts (and consequently, of the broader business environment, assuming that the panel constitutes a representative sample of the whole market). Another objective of this work was to come up with a conclusion about the "local" (or "global") nature of the proposed SOA MM.

At the end of this work, a new inter-enterprise SOA Maturity Model was developed and this was achieved by conducting a three-round questionnaire (as proposed by Delphi, which was used as the "vehicle" for this research) which was distributed to the panel experts who provided their feedback and contributed the most towards the new SOA MM realization.

P. Antoniades, *SOA Maturity Model*, SpringerBriefs in Information Systems,
DOI: 10.1007/978-3-319-02453-0_7, © The Author(s) 2014

In the proposed SOA maturity model (pSOAMM), five levels of SOA maturity were identified and a set of SOA domains along with their corresponding focus areas on which the domains would be evaluated, were defined. In general, the pSOAMM can be used by enterprises to facilitate their SOA adoption because pSOAMM provides an insight into the current SOA maturity level and a roadmap on how higher levels of SOA maturity can be achieved (and thus, benefit more from further SOA initiatives).

After building the new SOA maturity model (pSOAMM), it was used to capture the current SOA status and the SOA status in 5 years within the representative sample of local enterprises (through panel experts).

As far as the current SOA status is concerned, the majority of the panel experts' responses revealed that SOA is a relatively new concept to most of the experts' enterprises, but also that some of these enterprises have committed to adopting SOA even for certain parts of the organization. A smaller number of panel experts' enterprises have adopted SOA as a strategic enterprise-wide architectural paradigm.

Attitudes will be much more optimistic (in respect to SOA) in the next 5 years, according to the opinion of the representing body (panel experts) of the local market. In particular, panel experts believe that the SOA will be fundamental to the way the enterprises will operate and services might be used outside the enterprise, too (for instance, to be "consumed" by their business partners or customers). Moreover, they expect that their enterprises will operate a dynamic SOA with business and IT synchronized in order to achieve an optimum balance of agility, performance, risk and cost, in 5 years.

The research also revealed that there is an expressed trust (with an increasing trend) in SOA. This trust will be further boosted by repeated SOA projects' successes and by the proven benefits of SOA. Also, it can be concluded that SOA is gradually and steadily being accepted as the de-facto technology both in the IT and business community.

These findings comply with similar works implemented -mainly- in the US (InfoWorld 2006; CIO 2006), where the patterns revealed in this research also appear in those studies on a much larger scale and intensity, of course.

Interesting were also the findings in respect to the final objective of the research: to come up with a conclusion on whether the proposed SOA MM is applicable and useful to the local organizations, or whether it could be used (and provide business value) to organizations outside Cyprus, too. The conclusion was that even though the technology itself (SOA) is universal and some of the panel experts are working for global companies with operations in Cyprus, too, and most of them studied at foreign universities (so one may infer that the proposed SOA MM could be considered "global"), the fact that there was no participant from any other country (in particular, doing business in any other area of the world) "narrows" the proposed SOA MM "broadness" and that's why it is characterized as a "local" one, with no guarantee of its direct applicability in organizations outside Cyprus, but with the potential and completeness to be beneficial to other countries' organizations too, if appropriate (hopefully minor) adjustments are implemented.

Moreover, the research illustrated that IT services, SOA, SOA MM (and particularly the proposed SOA MM) seem to be ideal for Cyprus and any other country with similar business landscape (service-oriented) because of the size of the local market and of the nature of the business landscape (constant and rapid changes which require adaptive and agile technologies and enterprise architectural paradigms; SOA and its corresponding MMs seem to be an answer).

This work, though, has a number of limitations. Firstly, the proposed Delphi-derived inter-enterprise SOA MM was only applied to the organizations the experts came from. This constitutes a limited number of organizations in respect to the total. Future work may apply the newly proposed SOA Maturity Model in many more enterprises in the local market and elsewhere in order to further improve it, but also prove that the panel's opinions were a truly representative sample of the broader business landscape.

Limitation also constitutes the fact that the inter-enterprise "nature" of the proposed SOA MM was not heavily "tested". So, a future assignment would be to "stress" test the proposed SOA MM in order to fully prove its inter-enterprise "nature".

Moreover, the number of interviewees in the second group (which was composed of business owners and business practice managers) used for determining whether the proposed SOA MM is a "local" one (or, whether it could be used -with business value- by organizations outside Cyprus, too) was very limited (five) to gauge the opinion of the whole population on this issue. To obtain more accurate feedback, a greater number of interviews need to be conducted in the future.

Finally, the "locally" proposed SOA MM to be fully "promoted" into a "global" one, it has to be reviewed and adjusted accordingly by academia and industry representatives from other regions of the world too.

References

InfoWorld, (2006). Retrieved December, 2006, from http://www.infoworld.com.
CIO, (2006). Retrieved December, 2006, from http://www.cio.com.

Appendices

Appendix A: Delphi Technique Applied: First-Round Questionnaire

1. **Considering the various available SOA Maturity Models (distributed earlier to you) and the CMMI maturity model (also handed-out to you), how many levels/stages do you think it is advisable be included in a newly proposed SOA Maturity Model? [Select the one best reply]**
   a. Four, as proposed by the CBDI Forum
   b. Seven, as suggested by IBM
   c. Five, as described by HP's and Oracle's SOA Maturity Model
   d. Five, to also comply to CMMI model which, even though it is a Process Improvement Maturity Model, is considered to be the basis of all types/kinds of Maturity Models
   e. Other (please specify)
2. **If you have chosen (a) in the Question 1 above, what were the factors that made you choose that answer? [Select the one best reply]**
   a. The completeness of the SOA Maturity Model proposed by CBDI Form
   b. The high success rate of SOA projects by enterprises that adopted the CBDI Forum's SOA Maturity Model, according to independent studies/surveys
   c. CBDI Forum's credibility on SOA advising
   d. Other (please specify)
3. **If the answer provided for Question 1 was (b), why did you come up with such an answer? [Select the one best reply]**
   a. IBM is the most trust-worthy IT vendor and we would like to have IBM engaged in all IT- related activities (concerning Hardware, Software and Consulting Services) within our enterprise

P. Antoniades, *SOA Maturity Model*, SpringerBriefs in Information Systems, DOI: 10.1007/978-3-319-02453-0, © The Author(s) 2014

b. IBM's SOA Maturity Model seems to be the most complete
c. The high success rate of SOA projects by enterprises that adopted the IBM's SOA Maturity Model
d. Other (please specify)

4. **If you have provided (c) as your reply to Question 1 above, why have you made such an answer? [Select the one best reply]**

 a. The HP's and Oracle's Maturity Model is the most complete
 b. The high success rate of SOA projects by enterprises that adopted HP's and Oracle's SOA Maturity Model, according to independent studies/surveys
 c. Both HP and Oracle are among the most reliable IT vendors
 d. Other (please specify)

5. **If (d) was selected as an answer to Question 1 above, why have you made such a choice? [Select the one best reply]**

 a. CMMI was introduced by the leading IT academic institution in the US (Carnegie Mellon University) and as such it is recommended as the basis of all the Maturity Models of any type/kind
 b. The proposed SOA Maturity Model need not be based on any IT vendor's SOA Maturity Model(s), but instead on an independent's body/entity (if any)
 c. Other (please specify)

6. **Given the various SOA Maturity Models distributed earlier to you, which of the following domains/dimensions do you think might be considered as "SOA Enablers" and recommended for inclusion in the newly proposed SOA Maturity Model? [Select as many as you think are appropriate for inclusion in the proposed model]**

 a. Business / Organization
 b. Projects / Program Management
 c. Governance
 d. Architecture
 e. Enabling Technologies / IT Infrastructure
 f. Operations and Management
 g. Supply and Demand
 h. Delivery Methods
 i. Information or Data Format
 j. Business Processes
 k. People
 l. Other (please specify)

7. **If (a) was selected as an answer to Question 6 above, what are the three focus areas in which you are going to measure whether the domain/dimension (Business / Organization) has reached (or not) a certain level/stage of maturity? [Select the three best replies]**

 a. Whether (or not) there is and the degree of IT alignment with Business strategy

b. Change management (how easily the employees will adopt the change to SOA)
c. Whether (or not) there is and the degree of business involvement and understanding of SOA activities/projects
d. Whether (or not) there are and the degree to which SOA projects affect the organizational structure or management hierarchy
e. The level of impact of SOA projects on an organization's brand and reputation
f. Other (please specify)

8. **If you have chosen (b) in Question 6 above, what are the three focus areas in which you are going to measure whether the domain/dimension (Projects / Program Management) has reached (or not) a certain level/stage of maturity? [Select the three best replies]**

   a. Whether the SOA projects rollout is spanned across teams
   b. Whether the SOA projects rollout is spanned across departments
   c. Whether the SOA projects rollout is spanned across business units
   d. Whether the SOA projects rollout is spanned across the entire enterprise
   e. How well the management of the services portfolio is carried out
   f. Other (please specify)

9. **If the answer provided for Question 6 was (c), what are the three focus areas in which you are going to measure whether the domain/dimension (Governance) has reached (or not) a certain level/stage of maturity? [Select the three best replies]**

   a. Whether (or not) there is and the degree of funding for projects
   b. The scope (inter-enterprise, enterprise-wide, department-wide, etc) of organized SOA efforts
   c. Whether (or not) an enterprise is in discussions to acquire/merge with another organization
   d. Whether (or not) and to what extent there are policy enforcement schemes, reporting and exception handling guidelines/procedures across the collaborating enterprises
   e. Whether an enterprise is publicly or privately owned and managed
   f. Other (please specify)

10. **If you have provided (d) as your reply to Question 6 above, what are the three focus areas in which you are going to measure whether the domain/dimension (Architecture) has reached (or not) a certain level/stage of maturity? [Select the three best replies]**

   a. Whether there is an Enterprise Architecture (and which is its role) in the organization and whether it also governs the organization's relationships with its collaborating organizations
   b. Whether the Enterprise Architecture is being built by an organization's management or by its IT department

c. Whether (or not) there is and the degree to which services are cataloged and reused across the collaborating organizations
d. Whether the Enterprise Architecture describes the IT infrastructure (hardware, software, networking, etc) of an enterprise
e. Whether (or not) there is and the degree to which the Enterprise Architecture is being considered in the enterprise-wide and inter-enterprise planning activities
f. Other (please specify)

11. **If you have chosen (e) in Question 6 above, what are the three focus areas in which you are going to measure whether the domain/dimension (Enabling Technologies / IT Infrastructure) has reached (or not) a certain level/stage of maturity? [Select the three best replies]**

a. Whether the right tools (for software design, development, deployment, etc), technologies are in place
b. Whether industry standards are being followed (in respect to all aspects of IT operations)
c. Whether security, monitoring, and management tools (solutions) are in place
d. Whether the required (IT) expertise by human resources is found within the enterprise
e. Whether the .NET or the Java2 EE frameworks for software development and deployment are used
f. Other (please specify)

12. **If you have chosen (f) in Question 6 above, what are the three focus areas in which you are going to measure whether the domain/dimension (Operations and Management) has reached (or not) a certain level/stage of maturity? [Select the three best replies]**

a. Whether the right people exist within the enterprise to operate and manage the SOA activities/projects
b. Whether the people in charge of operating and managing the SOA activities/projects have sound knowledge and expertise in both IT and business
c. Whether the right procedures are in place to operate and manage the SOA activities/projects
d. Whether management and auditing tools (solutions) are in place
e. Whether any Quality-of-Service practices are followed
f. Other (please specify)

13. **If the answer provided for Question 6 was (g), what are the three focus areas in which you are going to measure whether the domain/dimension (Supply and Demand) has reached (or not) a certain level/stage of maturity? [Select the three best replies]**

a. The level of growth of SOA activities/projects within an enterprise
b. Whether any services' providers are out there in the industry with whom an enterprise might desire to co-operate
c. The degree of readiness of people (within an enterprise) to provide/use services

d. Whether any services' "consumers"/users are out there in the industry with whom an enterprise might desire to co-operate
e. The degree of readiness of an enterprise's lines-of-business or departments to use services
f. Other (please specify)

14. **If you have provided (h) as your reply to Question 6 above, what are the three focus areas in which you are going to measure whether the domain/dimension (Delivery Methods) has reached (or not) a certain level/stage of maturity? [Select the three best replies]**

a. Whether there are any policies, practices (throughout the collaborating enterprises) for delivering the SOA solutions
b. Whether .NET or Java2 EE-compliant platform is used as the deployment stage for SOA applications/solutions
c. Whether there are skills and expertise (throughout the collaborating enterprises) for delivering the SOA solutions
d. Whether there are any methodologies, modeling and abstraction techniques (throughout the collaborating enterprises) for delivering the SOA solutions
e. Whether there are dedicated people to deliver the SOA solutions
f. Other (please specify)

15. **If the answer provided for Question 6 was (i), what are the three focus areas in which you are going to measure whether the domain/dimension (Information or Data Format) has reached (or not) a certain level/stage of maturity? [Select the three best replies]**

a. Whether the collaborating enterprises are using industry standards for the representation of data (e.g. XML)
b. Whether there is any kind of auditing mechanism for checking who is accessing/modifying the enterprise data
c. Whether the data is in a single location (that is, whether there is any kind of data consolidation in place throughout the collaborating enterprises)
d. Whether (or not) the XBRL (eXtensible Business Reporting Language) format is supported
e. Whether the collaborating enterprises are not only using industry standards for the representation of data (like XML), but also other canonical formats and metadata management standards (like XSD, XSLT, XPath, Expression Language, etc)
f. Other (please specify)

16. **If you have chosen (j) in Question 6 above, what are the three focus areas in which you are going to measure whether the domain/dimension (Business Processes) has reached (or not) a certain level/stage of maturity? [Select the three best replies]**

a. Whether there is any kind of business process analysis toolkit used (within an enterprise)

b. Whether any kind of business process re-engineering has taken place lately in the enterprise
c. Whether there is any kind of business process monitoring tool used (throughout the collaborating enterprises)
d. Whether any kind of business process automation is in place
e. Whether a cross-application orchestration of business processes has been deployed (throughout the collaborating enterprises)
f. Other (please specify)

17. **If you have chosen (k) in Question 6 above, what are the three focus areas in which you are going to measure whether the domain/dimension (People) has reached (or not) a certain level/stage of maturity? [Select the three best replies]**

a. Whether the people (in an enterprise) have roles that are more of a combination of business and IT rather than being just one or the other
b. Whether there are constant organizational structure changes that are based around services
c. Whether an enterprise's human resources are shared across an enterprise and business managers
d. Whether an enterprise's Information Systems are not solely under a specific department's control, but is shared with others
e. Whether there is dependency of one department (of an enterprise) on other groups or other parts of the business
f. Other (please specify)

18. **Which of the following best describes your firm's approach to or status of SOA? [Select the one best reply]**

a. Not pursuing, and no immediate plans to do so
b. Will pursue within the next 12–24 months
c. Use selectively, without a clear strategy
d. Have an enterprise-level strategy and commitment for SOA
e. Other (please specify)

19. **What are you currently using SOA for?** [Select the one best reply]

a. Internal integration
b. External integration (that is, integration with partners' and/or customers' systems)
c. Strategic business transformation
d. Other (please specify)

20. **What are the most important factors that facilitate the "penetration" of the Service-Oriented Architecture and Web Services within your enterprise? [Select the three best replies]**

a. SOA (and Web Services) enhances agility and flexibility; important ingredients for our organization

b. SOA improves our products' or services' time-to-market
c. SOA reduces information systems integration complexity
d. SOA lowers costs by re-using existing IT assets
e. SOA creates new business opportunities through collaboration, outsourcing, and by enabling global presence for our goods and services
f. SOA accelerates growth by facilitating the integration of systems of acquired companies
g. Other (please specify)

21. **What are the primary constraints that prohibit the "penetration" of the Service-Oriented Architecture and Web Services within your enterprise? [Select the three best replies]**

    a. The technology (SOA) is still immature
    b. A lack of skills for the implementation and deployment of SOA-based ("Web Services-based") Information Systems
    c. There are still security implications on the technology
    d. Other (please specify)

22. **What will the SOA "penetration" be 5 years from now? [Select the one best reply]**

    a. SOA (and Web Services) will be the dominant technology 5 years from now, like what the "static" Web is today
    b. There will be a considerable (to great) increase in the adoption of SOA (and Web Services) by enterprises
    c. There will be a slight (or no) increase in the adoption of the technology
    d. It will be totally abandoned

23. **What will the status of SOA be 5 years from now, within your enterprise? [Select the one best reply]**

    a. Same answer as the one provided in the previous Question (22)
    b. Different answer as the one provided in the previous Question (22). If you provided this response as your reply to the current question, please explain why you had differentiated your enterprise from thc rest of the market's trend

24. **If you chose (a) or (b) in Question 22, what do you think will bc the most important facilitators toward that trend? [Select the one best reply]**

    a. There will be proven success stories associated with the application of SOA-based systems in highly demanding environments indicating the inherent benefits of the technology
    b. The ROI will be promising
    c. Greater business flexibility and agility
    d. Lowering of costs of integrating existing applications and systems
    e. Other (please specify)

25. **If you chose (c) or (d) in Question 22, what do you think will be the most important prohibiting factors for the adoption of the technology (SOA)? [Select the one best reply]**

   a. The ROI will never be encouraging enough
   b. The technology will fail to prove itself when it is implemented and deployed in systems where high security features are mandated
   c. The skills needed to implement and use the technology will be very limited in respect to the capacity required by the technology
   d. The Big supporters of the technology (Big IT companies like Microsoft, IBM, Oracle, etc) will abandon the technology and they will look for more profitable technologies/solutions
   e. Other (please specify)

Appendix B: Delphi Technique Applied: First-Round Questionnaire Findings

1. **Considering the various available SOA Maturity Models (distributed earlier to you) and the CMMI maturity model (also handed-out to you), how many levels/stages do you think might be included in a newly proposed SOA Maturity Model? [Select the one best reply]**

   a. Four, as proposed by the CBDI Forum [2/20 – 10 %]
   b. Seven, as suggested by IBM [6/20 – 30 %]
   c. Five, as described by HP's and Oracle's SOA Maturity Model [7/20 – 35 %]
   d. Five, to also comply to CMMI model which, even though it is a Process Improvement Maturity Model, is considered to be the basis of all types/kinds of Maturity Models [5/20 – 25 %]
   e. Other (please specify) [0/20 – 0 %]

2. **If you have chosen (a) in Question 1 above, what were the factors that made you choose that answer? [Select the one best reply]**

   a. The completeness of the SOA Maturity Model proposed by CBDI Form [0/2 – 0 %]
   b. The high success rate of SOA projects by enterprises that adopted the CBDI Forum's SOA Maturity Model, according to independent studies/surveys [1/2 – 50 %]
   c. CBDI Forum's credibility on SOA advising [1/2 – 50 %]
   d. Other (please specify) [0/2 – 0 %]

   Note: Two out of twenty respondents selected answer (a) in Question 1, above. That is why the "hit rate" in this Question was calculated based on two responses.

3. **If the answer chosen for Question 1 was (b), why did you come up with such an answer? [Select the one best reply]**

   a. IBM is the most trust-worthy IT vendor and we would like to have IBM engaged in all IT-related activities (concerning Hardware, Software and Consulting Services) within our enterprise [4/6 – 66.6 %]
   b. IBM's SOA Maturity Model seems to be the most complete [1/6 – 16.6 %]
   c. The high success rate of SOA projects by enterprises that adopted the IBM's SOA Maturity Model [1/6 – 16.6 %]
   d. Other (please specify) [0/6 – 0 %]

   Note: Six out of twenty respondents selected answer (b) in Question 1, above. That is why the "hit rate" in this Question was calculated based on six responses.

4. **If you have chosen (c) as your reply to Question 1 above, why have you made such an answer? [Select the one best reply]**

   a. The HP's and Oracle's Maturity Model is the most complete [4/7 – 57.1 %]
   b. The high success rate of SOA projects by enterprises that adopted HP's and Oracle's SOA Maturity Model, according to independent studies/surveys [2/7 – 28.6 %]
   c. Both HP and Oracle are among the most reliable IT vendors [1/7 – 14.3 %]
   d. Other (please specify) [0/7 – 0 %]

   Note: Seven out of twenty respondents selected answer (c) in Question 1, above. That is why the "hit rate" in this Question was calculated based on seven responses.

5. **If (d) was selected as an answer to Question 1 above, why have you made such a choice? [Select the one best reply]**

   a. CMMI was introduced by the leading IT academic institution in the US (Carnegie Mellon University) and as such it is recommended as the basis of all the Maturity Models of any type/kind [2/5 – 40 %]
   b. The proposed SOA Maturity Model need not be based on any IT vendor's SOA Maturity Model(s), but instead on an independent's body/entity model (if any) [3/5 – 60 %]
   c. Other (please specify) [0/5 – 0 %]

   Note: Five out of twenty respondents selected answer (d) in Question 1, above. That is why the "hit rate" in this Question was calculated based on five responses.

6. **Given the various SOA Maturity Models distributed earlier to you, which of the following domains/dimensions do you think might be considered as "SOA Enablers" and recommend be included in the newly proposed SOA Maturity Model? [Select as many as you think are appropriate to include in the proposed model]**

   a. Business / Organization [15]
   b. Projects / Program Management [5]
   c. Governance [15]
   d. Architecture [18]
   e. Enabling Technologies / IT Infrastructure [16]
   f. Operations and Management [6]
   g. Supply and Demand [4]
   h. Delivery Methods [15]
   i. Information or Data Format [17]
   j. Business Processes [18]
   k. People [9]
   l. Other (please specify) [0]

7. **If (a) was selected as an answer to Question 6 above, what are the three focus areas in which you are going to measure whether the domain/ dimension (Business / Organization) has reached (or not) a certain level/ stage of maturity? [Select the three best replies]**

   a. Whether (or not) there is and the degree of IT alignment with Business strategy [17/45 – 37.7 %]
   b. Change management (how easily the employees will adopt the change to SOA) [10/45 – 22.2 %]
   c. Whether (or not) there is and the degree of business involvement and understanding of SOA activities/projects [15/45 – 33.3 %]
   d. Whether (or not) there are effects and the degree to which SOA projects affect the organizational structure or management hierarchy [2/45 – 4.4 %]
   e. The level of impact of SOA projects on an organization's brand and reputation [1/45 – 2.2 %]
   f. Other (please specify) [0/45 – 0 %]

   Note: Fifteen out of twenty respondents selected answer (a) in Question 6 above. That is why the "hit rate" in this Question was calculated based on fifteen responses.
   In these types of questions, though, where the respondents were asked to choose three replies, the "hit rate" was calculated as if the number of respondents was 45 (15 participants X 3 replies).

8. **If you have chosen (b) in Question 6 above, what are the three focus areas in which you are going to measure whether the domain/dimension (Projects / Program Management) has reached (or not) a certain level/ stage of maturity?** [Select the three best replies]

   a. Whether the SOA projects rollout is spanned across teams [2/15 – 13.3 %]
   b. Whether the SOA projects rollout is spanned across departments [2/15 – 13.3 %]
   c. Whether the SOA projects rollout is spanned across business units [2/15 – 13.3 %]
   d. Whether the SOA projects rollout is spanned across the entire enterprise [7/15 – 46.6 %]
   e. How well the management of the services portfolio is carried out [2/15 – 13.3 %]
   f. Other (please specify) [0/15 – 0 %]

   Note: Five out of twenty respondents selected answer (b) in Question 6 above. That is why the "hit rate" in this Question was calculated based on five responses.
   In these types of questions, though, where the respondents were asked to choose three replies, the "hit rate" was calculated as if the number of respondents were 15 (5 participants X 3 replies).

9. **If the answer provided for Question 6 was (c), what are the three focus areas in which you are going to measure whether the domain/dimension (Governance) has reached (or not) a certain level/stage of maturity? [Select the three best replies]**

   a. Whether (or not) there is and the degree of funding for projects [13/45 – 28.9 %]
   b. The scope (inter-enterprise, enterprise-wide, department-wide, etc) of organized SOA efforts [15/45 – 33.3 %]
   c. Whether (or not) an enterprise is in discussions to acquire/merge with another organization [1/45 – 2.2 %]
   d. Whether (or not) and to what extent there are policy enforcement schemes, reporting and exception handling guidelines/procedures across the collaborating enterprises [15/45 – 33.3 %]
   e. Whether an enterprise is publicly or privately owned and managed [1/45 – 2.2 %]
   f. Other (please specify) [0/45 – 0 %]

   Note: Fifteen out of twenty respondents provided answer (c) in Question 6 above. That is why the "hit rate" in this Question was calculated based on fifteen responses.
   In these types of questions, though, where the respondents were asked to choose three replies, the "hit rate" was calculated as if the number of respondents was 45 (15 participants X 3 replies).

10. **If you have provided (d) as your reply to Question 6 above, what are the three focus areas on which you are going to measure whether the domain/dimension (Architecture) has reached (or not) a certain level/stage of maturity? [Select the three best replies]**

    a. Whether there is an Enterprise Architecture (and what is its role) in the organization and whether it also governs the organization's relationships with its collaborating organizations [16/54 – 29.6 %]
    b. Whether the Enterprise Architecture is being built by an organization's management or by its IT department [3/54 – 5.5 %]
    c. Whether (or not) there is and the degree to which services are cataloged and reused across the collaborating organizations [14/54 – 25.9 %]
    d. Whether the Enterprise Architecture describes the IT infrastructure (hardware, software, networking, etc) of an enterprise [6/54 – 11.1 %]
    e. Whether (or not) there is and the degree to which the Enterprise Architecture is being considered in the enterprise-wide and inter-enterprise planning activities [15/54 – 27.7 %]
    f. Other (please specify) [0/54 – 0 %]

    Note: Eighteen out of twenty respondents selected answer (d) in Question 6 above. That is why the "hit rate" in this Question was calculated based on eighteen responses.
    In these types of questions, though, where the respondents were asked to choose three replies, the "hit rate" was calculated as if the number of respondents was 54 (18 participants X 3 replies).

11. **If you have chosen (e) in Question 6 above, what are the three focus areas in which you are going to measure whether the domain/dimension (Enabling Technologies / IT Infrastructure) has reached (or not) a certain level/stage of maturity? [Select the three best replies]**

    a. Whether the right tools (for software design, development, deployment, etc), technologies are in place [14/48 – 29.1 %]
    b. Whether industry standards are being followed (in respect to all aspects of IT operations) [14/48 – 29.1 %]
    c. Whether security, monitoring, and management tools (solutions) are in place [14/48 – 29.1 %]
    d. Whether the required (IT) expertise by human resources (IT expertise) is found within the enterprise [3/48 – 6.3 %]
    e. Whether the .NET or the Java2 EE frameworks for software development and deployment are used [3/48 – 6.3 %]
    f. Other (please specify) [0/48 – 0 %]

    Note: Sixteen out of twenty respondents selected answer (e) in Question 6 above. That is why the "hit rate" in this Question was calculated based on sixteen responses.
    In these types of questions, though, where the respondents were asked to choose three replies, the "hit rate" was calculated as if the number of respondents was 48 (16 participants X 3 replies).

12. **If you have chosen (f) in Question 6 above, what are the three focus areas in which you are going to measure whether the domain/dimension (Operations and Management) has reached (or not) a certain level/stage of maturity? [Select the three best replies]**

    a. Whether the right people exist within the enterprise to operate and manage the SOA activities/projects [3/18 – 16.6 %]
    b. Whether the people in charge of operating and managing the SOA activities/projects have sound knowledge and expertise in both the IT and the business [6/18 – 33.3 %]
    c. Whether the right procedures are in place to operate and manage the SOA activities/projects [3/18 – 16.6 %]
    d. Whether management and auditing tools (solutions) are in place [3/18 – 16.6 %]
    e. Whether any Quality-of-Service practices are followed [3/18 – 16.6 %]
    f. Other (please specify) [0/18 – 0 %]

    Note: Six out of twenty respondents selected answer (f) in Question 6 above. That is why the "hit rate" in this Question was calculated based on six responses.
    In these types of questions, though, where the respondents were asked to choose three replies, the "hit rate" was calculated as if the number of respondents was 18 (6 participants X 3 replies).

13. **If the answer provided for Question 6 was (g), what are the three focus areas in which you are going to measure whether the domain/dimension (Supply and Demand) has reached (or not) a certain level/stage of maturity? [Select the three best replies]**

    a. The level of growth of SOA activities/projects within an enterprise [6/12 – 50 %]
    b. Whether any services' providers are out there in the industry with whom an enterprise might desire to co-operate [3/12 – 25 %]
    c. The degree of readiness of people (within an enterprise) to provide/use services [0/12 – 0 %]
    d. Whether any services' "consumers"/users are out there in the industry with whom an enterprise might desire to co-operate [3/12 – 25 %]
    e. The degree of readiness of an enterprise's lines-of-business or departments to use services [0/12 – 0 %]
    f. Other (please specify) [0/12 – 0 %]

    Note: Four out of twenty respondents selected answer (g) in Question 6 above. That is why the "hit rate" in this Question was calculated based on four responses.
    In these types of questions, though, where the respondents were asked to choose three replies, the "hit rate" was calculated as if the number of respondents was 12 (4 participants X 3 replies).

14. **If you have selected (h) as your reply to Question 6 above, what are the three focus areas in which you are going to measure whether the domain/dimension (Delivery Methods) has reached (or not) a certain level/stage of maturity? [Select the three best replies]**

    a. Whether there are any policies, practices (throughout the collaborating enterprises) for delivering the SOA solutions [15/45 – 33.3 %]
    b. Whether .NET or Java2 EE-compliant platform is used as the deployment stage for SOA applications/solutions [0/45 – 0 %]
    c. Whether there are skills and expertise (throughout the collaborating enterprises) for delivering the SOA solutions [15/45 – 33.3 %]
    d. Whether there are any methodologies, modeling and abstraction techniques (throughout the collaborating enterprises) for delivering the SOA solutions [15/45 – 33.3 %]
    e. Whether there are dedicated people for the delivery of the SOA solutions [0/45 – 0 %]
    f. Other (please specify) [0/45 – 0 %]

    Note: Fifteen out of twenty respondents selected answer (h) in Question 6, above. That is why the "hit rate" in this Question was calculated based on fifteen responses.
    In these types of questions, though, where the respondents were asked to choose three replies, the "hit rate" was calculated as if the number of respondents was 45 (15 participants X 3 replies).

15. **If the answer selected for Question 6 was the (i), what are the three focus areas in which you are going to measure whether the domain/dimension (Information or Data Format) has reached (or not) a certain level/stage of maturity? [Select the three best replies]**

    a. Whether the collaborating enterprises are using industry standards for the representation of data (e.g. XML) [18/51 – 35.3 %]
    b. Whether there is any kind of auditing mechanism for checking who is accessing/modifying the enterprise data [0/51 – 0 %]
    c. Whether the data is located on a single location (that is, whether there is any kind of data consolidation in place throughout the collaborating enterprises) [15/51 – 29.4 %]
    d. Whether (or not) the XBRL (eXtensible Business Reporting Language) format is supported [0/51 – 0 %]
    e. Whether the collaborating enterprises are not only using industry standards for the representation of data (like XML), but also other canonical formats and metadata management standards (like XSD, XSLT, XPath, Expression Language, etc) [18/51 – 35.3 %]
    f. Other (please specify) [0/51 – 0 %]

    Note: Seventeen out of twenty respondents selected answer (i) in Question 6 above. That is why the "hit rate" in this Question was calculated based on seventeen responses.
    In these types of questions, though, where the respondents were asked to choose three replies, the "hit rate" was calculated as if the number of respondents was 51 (17 participants X 3 replies).

16. **If you have chosen (j) in Question 6 above, what are the three focus areas in which you are going to measure whether the domain/dimension (Business Processes) has reached (or not) a certain level/stage of maturity? [Select the three best replies]**

    a. Whether there is any kind of business process analysis toolkit used (within an enterprise) [2/54 – 3.7 %]
    b. Whether any kind of business process re-engineering has taken place lately in the enterprise [2/54 – 3.7 %]
    c. Whether there is any kind of business process monitoring tool used (throughout the collaborating enterprises) [15/54 – 27.7 %]
    d. Whether any kind of business process automation is in place [20/54 – 37 %]
    e. Whether a cross-application orchestration of business processes has been deployed (throughout the collaborating enterprises) [15/54 – 27.7 %]
    f. Other (please specify) [0/54 – 0 %]

    Note: Eighteen out of twenty respondents selected answer (j) in Question 6 above. That is why the "hit rate" in this Question was calculated based on eighteen responses.
    In these types of questions, though, where the respondents were asked to choose three replies, the "hit rate" was calculated as if the number of respondents was 54 (18 participants X 3 replies).

17. **If you have chosen (k) in Question 6, what are the three focus areas in which you are going to measure whether the domain/dimension (People) has reached (or not) a certain level/stage of maturity? [Select the three best replies]**

    a. Whether the people (in an enterprise) have roles that are more of a combination of business and IT rather than being just one or the other [7/27 – 25.9 %]
    b. Whether there are constant organizational structure changes that are based around services [5/27 – 18.5 %]
    c. Whether an enterprise's human resources are shared across an enterprise and business managers [5/27 – 18.5 %]
    d. Whether an enterprise's Information Systems are not solely under a specific department's control, but they are shared with others [5/27 – 18.5 %]
    e. Whether there is dependency of one department (of an enterprise) on other groups or other parts of the business [5/27 – 18.5 %]
    f. Other (please specify) [0/27 – 0 %]

    Note: Nine out of twenty respondents selected answer (k) in Question 6. That is why the "hit rate" in this Question was calculated based on nine responses. In these types of questions, though, where the respondents were asked to choose three replies, the "hit rate" was calculated as if the number of respondents was 27 (9 participants X 3 replies).

18. **Which of the following best describes your firm's approach to or status of SOA? [Select the one best reply]**

    a. Not pursuing, and no immediate plans to do so [5/20 – 25 %]
    b. Will pursue within the next 12–24 months [5/20 – 25 %]
    c. Use selectively, without a clear strategy [6/20 – 30 %]
    d. Have an enterprise-level strategy and commitment for SOA [4/20 – 20 %]
    e. Other (please specify) [0/20 – 0 %]

19. **What are you currently using SOA for? [Select the one best reply]**

    a. Internal integration [5/10 – 50 %]
    b. External integration (that is, integration with partners' and/or customers' systems) [2/10 – 20 %]
    c. Strategic business transformation [3/10 – 30 %]
    d. Other (please specify) [0/10 – 0 %]

    Note: Ten out of twenty respondents selected answers (c) and (d) in Question 18 above. That is why the "hit rate" in this Question was calculated based on ten responses.

20. **Which are the most important factors that facilitate the "penetration" the Service-Oriented Architecture and Web Services within your enterprise? [Select the three best replies]**

    a. SOA (and Web Services) enhances agility and flexibility; important ingredients for our organization [20/60 – 33.3 %]
    b. SOA improves our products' or services' time-to-market [0/60 – 0 %]

c. SOA reduces information systems integration complexity [20/60 – 33.3 %]
d. SOA lowers costs by re-using existing IT assets [20/60 – 33.3 %]
e. SOA creates new business opportunities through collaboration, outsourcing, and by enabling global presence for our goods and services [0/60 – 0 %]
f. SOA accelerates growth by facilitating the integration of systems of acquired companies [0/60 – 0 %]
g. Other (please specify) [0/60 – 0 %]

Note: In these types of questions where the respondents were asked to choose three replies, the "hit rate" was calculated as if the number of respondents was 60 (20 participants X 3 replies).

21. **Which are the primary constraints that prohibit the "penetration" the Service-Oriented Architecture and Web Services within your enterprise? [Select the three best replies]**

a. The technology (SOA) is still immature [10/60 – 16.7 %]
b. There is a lack of skills for the implementation and deployment of SOA-based ("Web Services-based") Information Systems [20/60 – 33.3 %]
c. There are still security implications on the technology [0/60 – 0 %]
d. Other (please specify)
Lack of anticipated ROI [10/60 – 16.7 %]
Lack of funding [20/60 – 33.3 %]

Note: In these types of questions where the respondents were asked to choose three replies, the "hit rate" was calculated as if the number of respondents was 60 (20 participants X 3 replies).

22. **What will the SOA "penetration" be 5 years from now? [Select the one best reply]**

a. SOA (and Web Services) will be the dominant technology 5 years from now, like what the "static" Web is today [7/20 – 35 %]
b. There will be a considerable (to great) increase in the adoption of SOA (and Web Services) by enterprises [10/20 – 50 %]
c. There will be a slight (or no) increase in the adoption of the technology [3/20 – 15 %]
d. It will be totally abandoned [0/20 – 0 %]

23. **What will the status of SOA be 5 years from now, within your enterprise? [Select the one best reply]**

a. Same answer as the one provided in the previous Question (22) [20/20 – 100 %]
b. Different answer to the one provided in the previous Question (22). If you had provided this response as your reply to the current question, please explain why you had differentiated your enterprise from the rest of the market's trend [0/20 – 0 %]

24. **If you had selected answers (a) or (b) in Question 22, which do you think will be the most important facilitators toward that trend? [Select the three best replies]**

    a. There will be proven success stories associated with the application of SOA-based systems in highly demanding environments indicating the inherent benefits of the technology [10/51 – 19.6 %]
    b. The ROI will be promising [7/51 – 13.7 %]
    c. Greater business flexibility and agility [17/51 – 33.3 %]
    d. Lowering of costs of integrating existing applications and systems [17/51 – 33.3 %]
    e. Other (please specify) [0/51 – 0 %]

    Note: Seventeen out of twenty respondents selected answer (a) or (b) in Question 22 above. That is why the "hit rate" in this Question was calculated based on seventeen responses.
    However, based on what has been done in these types of questions earlier, the "hit rate" was calculated as if the number of respondents was 51 (17 participants X 3 replies).

25. **If you had selected answers (c) or (d) in Question 22, which do you think will be the most important prohibiting factors for the adoption of the technology (SOA)? [Select the three best replies]**

    a. The ROI will never be encouraging enough [2/9 – 22.2 %]
    b. The technology will fail to prove itself when it is implemented and deployed in systems where high security features are mandated [2/9 – 22.2 %]
    c. The skills needed to implement and use the technology will be very limited in respect to the capacity required by the technology [4/9 – 44.4 %]
    d. The Big supporters of the technology (Big IT companies like Microsoft, IBM, Oracle, etc) will abandon the technology and they will look for more profitable technologies/solutions [1/9 – 11.1 %]
    e. Other (please specify)

    Note: Only three out of twenty respondents provided answers (c) and (d) in Question 22 above. That is why the "hit rate" in this Question was calculated based on three responses.
    However, based on what has been done in these types of questions earlier, the "hit rate" was calculated as if the number of respondents was 9 (3 participants X 3 replies).

Appendix C: Delphi Technique Applied: Second-Round Questionnaire

1. **Based on the first-round questionnaire handed-out,**

    a. 60 % of the respondents suggested that the proposed SOA Maturity Model might have five levels/stages, as described by HP's and Oracle's SOA Maturity Model (35 %) and in order to comply to CMMI Model which is considered the basis of all types/kinds of Maturity Models (25 %)
    b. 30 % of the respondents proposed seven levels/stages as recommended by IBM
    c. 10 % of the respondents selected four, as proposed by CBDI Forum

    when asked: "how many levels/stages do you think might be included in a newly proposed SOA Maturity Model?"
    Knowning the preferences of the respondents(above), what do you think of the above statements? Which one from (a) to (c) is more likely to be closer to what is ideal to include in the proposed SOA Maturity Model?

    This also implies that not only will the number of levels/stages of the proposed SOA Maturity Model be the same as those of HP's and Oracle's and CMMI's, but also the interpretation provided by these models for each of these levels/stages (that is, Level 1 means that SOA is a relatively new concept in the enterprise, etc) will resemble that of the proposed model.

    Do you agree with this statement?

    i. Yes
    ii. No.

    If you have chosen "No", please indicate any other interpretation.

2. **The findings of the first-round questionnaire revealed that 60 % of the respondents suggested that the proposed SOA Maturity Model might have five levels/stages, as described by HP's and Oracle's SOA Maturity Model (35 %) and in order to comply with CMMI Model, which is considered the basis of all types/kinds of Maturity Models (25 %). Out of those who responded in favor of five levels/stages because of HP's and Oracle's SOA Maturity Model:**

    a1. 57.1 % of them have chosen to do so because HP's and Oracle's Maturity Model is the most complete
    b1. 28.6 % of them because of the high success rate of SOA projects by enterprises that adopted HP's and Oracle's SOA Maturity Model, according to independent studies/surveys
    c1. 14.3 % of them because both HP and Oracle are among the most reliable IT vendors

Also, out of those who selected five levels/stages because they wanted the proposed SOA Maturity Model to comply with CMMI Model which is considered the basis of all types/kinds of Maturity Models:

a2. 60 % of them have supported their preference on the fact that "the proposed SOA Maturity Model need not be based on any IT vendor's SOA Maturity Model(s), but instead on an independent's body/entity model (if any)"
b2. 40 % of them because CMMI was introduced by the leading IT academic institution in the US (Carnegie Mellon University) and as such it is recommended as the basis of all the Maturity Models of any type/kind

when asked to justify their preference to a SOA Maturity Model with five levels/stages.
Considering these replies, which of them do you think can be used as strong arguments in favor of five levels/stages? (Choose from a1 to c1, and a2 or b2, above).

3. **In the first-round questionnaire, looking for the domains/dimensions that might be considered as "SOA Enablers" and are recommended be included in the newly proposed SOA Maturity Model:**

   a. 18/20 respondents were in favor of "Architecture" and "Business Processes"
   b. 17/20 respondents selected "Information or Data Format"
   c. 16/20 respondents selected "Enabling Technologies / IT Infrastructure"
   d. 15/20 respondents selected "Governance", "Business / Organization" and "Delivery Methods"

   Having this piece of information in mind, do you agree that these are the most important domains/dimensions that might be included in the newly proposed SOA Maturity Model?

   i. Yes
   ii. No.

   If you choose to go with the "No", please specify any other domains/ dimensions you think might be included in the model.

4. **The following is the distribution of the percentages of the respondents who chose "Business / Organization" as one of the domains/dimensions that need to be included in the proposed model and who were called to identify the three focus areas in which this specific domain would be evaluated:**

   a. 37.7 % of respondents selected "Whether (or not) there is and the degree of IT alignment with Business strategy"
   b. 33.3 % of respondents selected "Whether (or not) there is and the degree of business involvement and understanding of SOA activities/projects"
   c. 22.2 % of respondents selected "Change management (how easily the employees will adopt the change to SOA)"

d. 4.4 % of respondents selected "Whether (or not) there are and the degree to which SOA projects affect the organizational structure or management hierarchy"
e. 2.2 % of respondents selected "The level of impact of SOA projects on an organization's brand and reputation"

With which three of the statements above, are you more likely agree with as the most appropriate to be included as focus areas for this domain?

5. **Considering the results of the first-round questionnaire,**

a. 33.3 % of the participants have chosen "The scope (inter-enterprise, enterprise-wide, department-wide, etc) of organized SOA efforts" and
b. An equal percent of the participants (33.3) the "Whether (or not) and to what extent there are policy enforcement schemes, reporting and exception handling guidelines/procedures across the collaborating enterprises"
c. 28.9 %, the "Whether (or not) there is and the degree of funding for projects"
d. 2.2 %, the "Whether (or not) an enterprise is in discussions for acquiring/ merging with another organization" and
e. Another 2.2 % the "Whether an enterprise is publicly or privately owned and managed"

as the most appropriate (three) focus areas on which the "Governance" domain/dimension would be evaluated. With which three options (from –a- to –e-) do you agree more?

6. **The first-round questionnaire demonstrated that:**

a. 29.6 % of the respondents selected the "Whether there is an Enterprise Architecture (and what is its role) in the organization and whether it also governs the organization's relationships with its collaborating organizations"
b. 27.7 %, the "Whether (or not) there is and the degree to which the Enterprise Architecture is being considered in the enterprise-wide and inter-enterprise planning activities"
c. 25.9 %, the "Whether (or not) there is and the degree to which services are cataloged and reused across the collaborating organizations"
d. 11.1 %, the "Whether the Enterprise Architecture describes the IT infrastructure (hardware, software, networking, etc) of an enterprise"
e. 5.5 %, the "Whether the Enterprise Architecture is being built by an organization's management or by its IT department"

as the possible focus areas for the "Architecture" domain/dimension. Based on this piece of information, what do you think? Which of these options do you believe are the three most important focus areas to be included in the proposed SOA Maturity Model?

7. **The population distribution of responses provided for the candidate focus areas concerning the "Enabling Technologies / IT Infrastructure" domain/dimension were:**

    a. 29.1 % of the participants replied that the focus area: "Whether the right tools (for software design, development, deployment, etc) and technologies are in place" might be included as a metric for the "Enabling Technologies / IT Infrastructure" domain
    b. 29.1 % selected "Whether industry standards are being followed (in respect to all aspects of IT operations)"
    c. 29.1 % selected "Whether security, monitoring, and management tools (solutions) are in place"
    d. 6.3 % selected "Whether the required (IT) expertise by human resources is found within the enterprise"
    e. 6.3 % selected "Whether .NET or Java2 EE frameworks for software development and deployment are used"

    Which three from (a) to (e) above do you qualify to be used as focus areas concerning the "Enabling Technologies / IT Infrastructure" domain?

8. **Looking for the three most appropriate focus areas in which the domain/ dimension "Delivery Methods" would be evaluated, the following responses were obtained from the first-round questionnaire:**

    a. 33.3 % of the respondents selected "Whether there are any policies and practices (throughout the collaborating enterprises) for delivering the SOA solutions"
    b. 33.3 % of the respondents selected "Whether there are skills and expertise (throughout the collaborating enterprises) for delivering the SOA solutions"
    c. 33.3 % of the respondents selected "Whether there are any methodologies, modeling and abstraction techniques (throughout the collaborating enterprises) for delivering the SOA solutions"

    Do you agree that these three options are the most appropriate focus areas for evaluating the "Delivery Methods" domain?

    i. Yes
    ii. No.

    If you choose to go with the "No", please specify any other focus areas you think might be considered for the "Delivery Methods" domain.

9. **The replies to the question requesting the three most appropriate focus areas for evaluating the "Information or Data Format" domain were:**

    a. 35.3 % of respondents selected "Whether the collaborating enterprises are using industry standards for the representation of data (e.g. XML)"
    b. 35.3 % of respondents selected "Whether the collaborating enterprises are not only using industry standards for the representation of data (like

XML), but also other canonical formats and metadata management standards (like XSD, XSLT, XPath, Expression Language, etc)"
c. 29.4 % of respondents selected "Whether the data is located on a single location (that is, whether there is any kind of data consolidation in place throughout the collaborating enterprises)"

Is that what you expected?

i. Yes
ii. No.

If you choose to go with the "No", please specify any other focus areas you think might be considered for the "Information or Data Format" domain.

10. **The population distribution of responses provided as candidate focus areas concerning the "Business Processes" domain/dimension were:**

    a. 37 % of the respondents chose: "Whether any kind of business process automation is in place"
    b. 27.7 % of the respondents chose: "Whether a cross-application orchestration of business processes has been deployed (throughout the collaborating enterprises)"
    c. 27.7 % of the respondents chose: "Whether there is any kind of business process monitoring tool used (throughout the collaborating enterprises)"
    d. 3.7 % of the respondents chose: "Whether there is any kind of business process analysis toolkit used (within an enterprise)"
    e. 3.7 % of the respondents chose: "Whether any kind of business process re-engineering has taken place lately in the enterprise"

    Which three from (a) to (e) above do you qualify to be used as focus areas for evaluating the "Business Process" domain?

11. **The first-round questionnaire also revealed information concerning the panel members' enterprises status in respect to SOA. In particular:**

    a. 50 % of the respondents are currently using SOA (either selectively, without a clear strategy −30 %-, or they do "have an enterprise-level strategy and commitment for SOA" −20 %-)
    b. 25 % of the respondents "will pursue within the next 12–24 months"
    c. 25 % of the respondents are "not pursuing, and no immediate plans to do so"

    Do you think that these responses properly illustrate the SOA status in your enterprise (and subsequently, in the local market)?

    i. Yes
    ii. No.

    If you choose to go with the "No", please specify what you think is a more appropriate statement for describing the SOA status in your organization (and, subsequently, in the country).

12. **As far as the SOA applicability (that is, in which areas or type of projects SOA is being used) is concerned, the participants' replies were:**

    a. 50 % of the respondents selected "Internal integration"
    b. 30 % selected "Strategic business transformation"
    c. 20 % selected "External integration (that is, integration with partners' and/or customers' systems)"

    Do you think that these are the areas or types of projects SOA is mainly being used for?

    i. Yes
    ii. No.

    If you choose to go with the "No", please specify what other areas or types of projects SOA might be used for.

13. **According to the first-round questionnaire's findings, the most important factors that facilitate the "penetration" of Service-Oriented Architecture and Web Services within IT experts' enterprises are:**

    a. "SOA (and Web Services) enhances agility and flexibility; important ingredients for our organization" (33.3 %)
    b. "SOA reduces information systems integration complexity" (33.3 %),
    c. "SOA lowers costs by re-using existing IT assets" (33.3 %)

    It is widely accepted that all three factors chosen (above) are the most important facilitators for SOA "penetration" within enterprises. The same piece of information has been revealed by the findings of the first-round questionnaire. Is that acceptable to you?

    i. Yes
    ii. No.

    If you choose to go with the "No", please specify any other important facilitators for SOA "penetration" within your organization.

14. **It was also found (in the first-round questionnaire) that the primary constraints that prohibit the "penetration" of Service-Oriented Architecture and Web Services within enterprises are:**

    a. "Lack of funding" (33.3 %)
    b. "There is a lack of skills for the implementation and deployment of SOA-based ("Web Services-based") Information Systems" (33.3 %),
    c. "Lack of anticipated ROI" (16.7 %)
    d. "The technology (SOA) is still immature" (16.7 %),

    Do you agree with these findings?

    i. Yes
    ii. No.

If you choose to go with the "No", please specify any other important prohibiting factor(s) for SOA "penetration" within your organization.

15. **Another finding derived from the first-round questionnaire was that:**

    a. "There will be a considerable (to great) increase in the adoption of SOA (and Web Services) by enterprises" (50 %),
    b. "SOA (and Web Services) will be the dominant technology 5 years from now, like what the -static- Web is today" (35 %),
    c. "There will be a slight (or no) increase in the adoption of the technology" (15 %),

    when the IT experts were asked to "predict" the SOA penetration 5 years from now, both within their enterprises and to the broader business landscape.

    Is that what you expected?

    i. Yes
    ii. No.

    If you choose to go with the "No", please specify what you anticipate will happen in 5 years, in respect to SOA.

16. **In addition to the above "prediction" (Question 15), the IT experts provided their opinions regarding which factors they consider most important toward facilitating the adoption of SOA (and Web Services) by enterprises 5 years from today.**

    a. 33.3 % of them believe that SOA provides "Greater business flexibility and agility",
    b. 33.3 %, "Lowering of costs of integrating existing applications and systems",
    c. 19.6 %, "There will be proven success stories associated with the application of SOA-based systems in highly demanding environments indicating the inherent benefits of the technology", and
    d. 13.7 %, "The ROI will be promising".

    Do you agree with these findings?

    i. Yes
    ii. No.

    If you choose to go with the "No", please specify any other important facilitator(s) for SOA "penetration" within enterprises in 5 years' time.

17. **Moreover, the IT experts provided their opinions regarding which factors they consider as most prohibiting towards the adoption of SOA (and Web Services) by enterprises, 5 years from today.**

    a. 44.4 % believe that "The skills needed to implement and use the technology will be very limited in respect to the capacity required by the technology",
    b. 22.2 %, "The ROI will never be encouraging enough",

c. 22.2 %, "The technology will fail to prove itself when it is implemented and deployed in systems where high security features are mandated",
d. 11.1 %, "The big supporters of the technology (Big IT companies like Microsoft, IBM, racle, etc) will abandon the technology and they will look for more profitable technologies/solutions".

Do you think that these responses are what you were expecting?

i. Yes
ii. No.

If you choose to go with the "No", please specify any other important prohibiting factor(s) for SOA "penetration" within enterprises in 5 years' time.

Appendix D: Delphi Technique Applied: Third-Round Questionnaire

The second-round questionnaire revealed that three-quarters (that is, 75 %) of the panel members are currently using SOA (either selectively, without a clear strategy or they do "have an enterprise-level strategy and commitment to SOA") or they will pursue SOA projects within the next 12–24 months.

The remaining quarter (25 %) of the respondents admitted that they are "not pursuing, and no immediate plans to do so."

As far as the future status (in particular, after 5 years) of SOA is concerned, a very large number of respondents replied that "there will be a considerable (to great) increase in the adoption of SOA (and Web Services)" and that "SOA (and Web Services) will be the dominant technology 5 years from now, like what the-static-Web is today", demonstrating not only their willingness to "embrace" the technology (SOA), but also the technology's reliability over the last few years and its increasing acceptance in the business.

1. Considering:
    i. this piece of information stated above,
    ii. the information you had provided in the two (2) questionnaires distributed to you earlier,
    iii. what is really happening within your organizations in respect to SOA, and,
    iv. the derived SOA Maturity Model which was the outcome of the second-round questionnaire (and it was handed-out to you along with this questionnaire),where would you locate the current status of your enterprise (that is, into which maturity level) in respect to the various SOA domains identified in the model?Please designate your choices by drawing the corresponding boxes in the SOA Maturity Model (Table 5.1) which was handed out to you.
2. Repeat what has been done in Question 1 above, but now indicate (if possible, using a different color) the status of SOA in 5 years' time.

Appendix E: Delphi Technique Applied: Third-Round Questionnaire Sample Responses

E1. Current Status Of SOA:

Panel Member X—Sample Responses:

	Maturity Levels				
	Level 1	Level 2	Level 3	Level 4	Level 5
pSOAMM's SOA Enabler / Domain (Along With Its Focus Areas)					
Architecture (*Focus Areas:* a. Role of Enterprise Architecture; b. Cataloging and Reuse; c. Planning and Guidance)					
Infrastructure (*Focus Areas:* a. Standards; b. Security and Monitoring; c. Management and Operations)					

Delivery (*Focus Areas:* a. Project vs. Enterprise Focus; b. Skills and Methodologies in Place; c. Modeling and Abstraction Techniques)					
Information (*Focus Areas:* a. Data Standards and Canonical Formats; b. Metadata Management; c. Single Source of Truth)					
Process (*Focus Areas*: a. Process Automation; b. Composite Application					

Development; c. Process Measurement and Scoring)					
Organization (*Focus Areas:* a. IT Alignment with Business Strategy; b. Change Management; c. Business Involvement and Understanding)					
Governance (*Focus Areas:* a. Funding and Accounting; b. Inter-Enterprise / Cross-Organizational Involvement; c. Policies, Reporting, and Exception Handling)					

E2. Status Of SOA In 5 Years

Panel Member X—Sample Responses:

	Maturity Levels				
	Level 1	Level 2	Level 3	Level 4	Level 5
pSOAMM's SOA Enabler / Domain (Along With Its Focus Areas)					
Architecture (*Focus Areas:* a. Role of Enterprise Architecture; b. Cataloging and Reuse; c. Planning and Guidance)					

Infrastructure (*Focus Areas:* a. Standards; b. Security and Monitoring; c. Management and Operations)					
Delivery (*Focus Areas:* a. Project vs. Enterprise Focus; b. Skills and Methodologies in Place; c. Modeling and Abstraction Techniques)					
Information (*Focus Areas:* a. Data Standards and Canonical					

Formats; b. Metadata Management; c. Single Source of Truth)					
Process (*Focus Areas*: a. Process Automation; b. Composite Application Development; c. Process Measurement and Scoring)					
Organization (*Focus Areas:* a. IT Alignment with Business Strategy; b. Change Management;					

c. Business Involvement and Understanding)					
Governance (*Focus Areas:* a. Funding and Accounting; b. Inter-Enterprise / Cross-Organizational Involvement; c. Policies, Reporting, and Exception Handling)					

Appendix F: Selected Panel Experts Interview (On Proposed SOA MM "Locality")

Interview Agenda

Setting

Objective of Interview:

Determine whether the proposed SOA MM can only be used locally or whether it is a "global" model.

Interview

Points To Be Discussed:

1. Based on the brief introduction handed-out to you regarding the Delphi technique (as shown in Sect. 4.1 *and which was used as the "vehicle" for carrying out this research work and allowing me to come up with the proposed SOA), do you think that this technique is the most appropriate tool for carrying out this type of research work (proposing a new SOA MM)?*
2. *Do you think that the proposed SOA MM (which has been derived through your contribution, too) is for local organizations only or can it be used (and have a real value) by organizations outside Cyprus, too?*
3. *If you think that the proposed SOA MM is only for Cypriot organizations, what are those factors that make you feel confident of your opinion / perception?*
4. *On the other hand, if you think that the proposed SOA MM can be used by organizations outside Cyprus (that is, it is a "global" model), what are those ingredients that make it "global"?*
5. *What are those domains/factors that could be added to the proposed SOA MM that would make it "global"?*
6. *Is there anything else that could be added to the proposed SOA MM to make it even more complete, broad and "global" one?*

Follow-Up

There was no need to have a follow-up session with any of the selected panel experts.

Appendix G: The Business Imperative For SOA (Integration)

Executive Summary

While change may have been a constant since time immemorial, the rate of change is accelerating far faster than ever before, and this is having a profound effect on business. Business cycles are shrinking rapidly. The way business was conducted even a decade ago is no longer acceptable if an organization wishes to remain competitive. Organizations have had to change how they interact with customers, how they manufacture goods, and how they are organized and managed.

Rapid changes are only possible when the organization itself is agile. The notion of an agile business has long captured the imagination of business executives. The agile business is able to embrace changes in market conditions, organizational structure, and the regulatory environment without missing a beat. An agile business empowers the management team to focus its collective acumen on delivering substantially increased value to its stakeholders.

The desire for greater business agility is making integration increasingly important. Agility is the combination of speed and adaptability: speed to bring new solutions to markets more quickly, and adaptability to new business requirements and competitive pressures. This can only be achieved when business processes are able to change easily. And that, in turn, is possible only when the underlying IT systems are integrated flexibly to accommodate the speed of change required.

Integration also enables organizations to leverage existing IT investments to streamline their processes for greater efficiency and productivity. IT infrastructures need to utilize existing applications and systems to the extent possible. This requires an integrated infrastructure along with end-to-end visibility of the business processes across disparate systems. It also requires an approach that supports the implementation of business solutions from existing components. This is the idea behind Service Oriented Architecture, or SOA.

Challenges to Business Agility

Currently, business agility is constrained by a number of obstacles. These include inflexible applications that cannot be easily changed or enhanced; non-integrated stovepipe applications; inefficient business processes; a lack of visibility into business processes and operations; challenges brought about by mergers, acquisitions, and regulations; and a lack of alignment between an organization's strategic objectives and daily operations.

Integration technology can play a role in removing these obstacles. In fact, in today's business climate, organizational growth may be inhibited without integration. Integration technology is a key enabling factor in helping business and IT executives transform their organizations; get to market more quickly; respond faster to business opportunities, competitive pressures, and regulatory requirements; and differentiate how their organizations do business. Business responsiveness has become a function of an organization's ability to rapidly marshal the underlying IT systems in alignment with business needs. This means leveraging existing assets while creating new business functionality.

The Mandate for Integration

One only needs to look at the history of computing and the evolution of business software to see why integration is a priority for most chief information officers (CIOs) today. In the beginning, back-office operations were run by mainframe systems. These systems were built to optimize expensive computing resources, but not business agility. They were—and remain—difficult to change and brittle to boot: Changing one thing could easily break something else. The legacy systems still in place today generally run an organization's core back-end operations in a robust and reliable way. However, they are usually batch systems and are unable to respond flexibly to business needs for realtime information or new functionality. That unfulfilled business need gave rise to the first wave of distributed computing with the emergence of minicomputers. Department managers could purchase them with their budgets and select them based on their particular needs. Often, the availability of a packaged application that met 50–80 % of the department's specific needs led to the introduction of these new platforms.

Unix, PCs, and client server software further reduced the cost of department systems, and they began to proliferate throughout organizations. Low-cost desktop productivity and development tools—such as Microsoft Excel, Access, and Visual Basic—supported ad hoc solutions, which also proliferated and often became more strategic than they were originally designed to be. Individual business units established their own computing facilities and application development capabilities, in effect setting up shadow IT organizations. This gave them the desired level of agility—at least at first—to respond more rapidly and independently to new business needs. However, it also led to islands of automation: hundreds of applications spread across the organization, many of them on desktops. These applications used organizational information and were in turn used to report on business operations. But they were not under organizational IT management and were certainly not integrated consistently.

The emergence of the Internet and the associated rush to e-business further punctuated the need for organizations—and, by association, IT—to become more responsive to market dynamics. But with different departments launching their own initiatives and often duplicating one another's work, the result was reduced visibility and control, along with reduced economies of scale.

The widespread adoption of distributed systems often meant that large organizations had multiple platforms running hundreds of applications that managed similar information through different portions of various business processes. Unfortunately, these applications were not designed to integrate with one another, so the organizations had to find ways to keep the information in sync across the systems. Rekeying the data—a means of last resort, but all too often the approach taken—was slow, resource intensive, and prone to errors that could be costly to trace and resolve. So an easier way to integrate the disparate stand-alone systems was needed.

Another problem was that each packaged application was designed to focus on specific department processes. Business agility requires the optimization of business processes end-to-end. This includes improving the process for initiators such as customers, partners, and suppliers as well as people in different business roles, and developing systems that support different parts of the business.

While some emerging technologies look for a business problem to address, integration was a business problem long before it was a technology. Packaged software systems were far from being turnkey solutions. The integration costs of implementing the typical enterprise resource planning (ERP) system could be three to five times the cost of the software. The problem was that the integration involved point-to-point hand-coding. This required an understanding of the application program interfaces (APIs) of all the systems to be integrated and a high level of expertise. It also took a great deal of time and was inflexible to change. The number of interfaces rose exponentially with the number of systems being integrated. Such integration "spaghetti" was difficult to manage and change, and upgrading to a new version of any of the applications meant going through the process all over again.

This brings us to the present. Although the current state of affairs is largely the result of short-term business decisions, it is ultimately viewed as an IT problem. That's because business change is intimately tied to the underlying IT systems' flexibility—or lack thereof. As markets move quickly and new opportunities and competitors emerge, organizations are increasingly challenged to close the gap between their business needs and the lack of flexibility in their IT infrastructures. Perhaps the biggest obstacle to closing this gap is that change has to be implemented while the core IT operations that support the business continue to function smoothly and seamlessly. The challenge is akin to upgrading the wings of a plane while it's in flight. This is where Service Oriented Architecture enters the picture.

Integration and SOA

At the core, SOA is about creating systems out of standard building blocks. The concept is not new. Many of the principles underlying SOA, such as isolating functionality to promote reusability, are long-standing best practices. In the past, however, the adoption of SOA was hindered by a lack of widely accepted standards for putting together these building blocks. Now, the pressing demands of business—which necessitate increased system flexibility and adaptability—have inspired nearly all organizations to align on a single set of standards for SOA: Web services.

While SOA involves much more than Web services, standardization is largely responsible for removing the barriers to SOA and fostering a future where systems can be more easily assembled and incrementally modified. Today, in the face of growing competitive pressure and the accelerating pace of business, organizations realize that they run a risk if they do not move toward SOA.

So how are SOA and integration related? In fact, they are highly complementary.

SOA is inherently about a distributed architecture, with systems that span computing platforms, data sources, and technologies. A distributed architecture requires integration. By standardizing how systems interoperate, Web services simplify the task of integration. Web services alone, however, do not suffice. Organizations need an evolutionary approach to SOA that incorporates legacy (non-Web-services-based) systems. Integration software provides the bridge between the legacy systems and SOA, allowing organizations to leverage existing software assets while managing their transition to SOA.

Integration solutions also contribute mature technologies—such as messaging, routing, data translation and transformation, and event management—along with organizational disciplines that are necessary for full-fledged, enterprise SOA. Moreover, integration capabilities such as business process management (BPM) and business activity monitoring (BAM) allow organizations to realize a higher level of business productivity from SOA by enabling the optimization of business processes and the alignment of strategic objectives with operational actions. In short, integration should play a central role in any organization's SOA strategy.

Appendix H: Business Practitioners (Business Owners And Business Practice Managers) Interview (On Proposed SOA MM "Locality")

Interview agenda
Setting
Objective of Interview:
Determine whether the proposed SOA MM can only be used locally or whether it is a "global" model..
Interview
Points To Be Discussed:
1. *The proposed SOA MM (handed-out to you, along with an Introduction on SOA titled: "The Business Imperative For SOA"- shown as Appendix G-, and an introduction on Delphi technique (-as shown in* Sect. 4.1*-) evolved from the contribution/feedback provided by of a team of experts (both from Academia and Industry) and it aims at: (a) helping you position your company's initiatives in respect to SOA initiatives, (b) anticipating what that status (in respect to SOA) might be in a few years, and (c) providing guidance for achieving higher levels of SOA maturity within your organizations. Do you think that the proposed SOA MM takes into consideration real, day-to-day business challenges, or it is just a theoretical framework with no (or minimal) business value?*
2. *Based on the brief introduction handed-out to you regarding the Delphi technique (which had been used as the "vehicle" for carrying out this research work and allowing me to come up with the proposed SOA), do you think that this technique is the most appropriate tool for carrying out this type of research work (proposing a new SOA MM)?*
3. *Do you think that the proposed SOA MM addresses the challenge of incorporating those domains/factors/"ingredients" that refer to inter-enterprise settings?*
4. *Do you think that the proposed SOA MM is geared towards local organizations only or can it be used (and have a real value) by organizations outside Cyprus, too?*
5. *If you think that the proposed SOA MM is appropriate only for Cypriot organizations, what are the factors that make you feel confident of your opinion / perception?*
6. *On the other hand, if you think that the proposed SOA MM can be used by organizations outside Cyprus (that is, it is a "global" model), what are the ingredients that make it "global"?*
7. *What are the domains/factors that could be added on the proposed SOA MM that would have made it "global"?*
8. *Is there anything else that could be added to the proposed SOA MM to make it even more complete, broad and "global"?*
Follow-Up
I had to insist a bit on the interviews with business practitioners (business owners and business practice managers) because –most of the time- they were busy;
I also had to "educate" them on concepts/notions like SOA, Maturity Models, Delphi, etc, in order to acquire/"capture" the information required. So, a number of follow-up sessions took place with this group of interviewees.

Appendix I: Major SOA Maturity Models

CBDI Forum's SOA Maturity Model

CBDI Forum, a "dedicated-to-SOA" research organization, advises that "in order to synchronize business requirements, technology capabilities and investments with ROI, a roadmap approach which identifies stages of maturity for the various areas impacted, is needed." (CBDI 2006) In particular, Fig. I.1 below shows a generic SOA Roadmap which links SOA drivers and strategies to stages of maturity, as proposed by CBDI (2006).

In the SOA Roadmap proposed by CBDI Forum, four(4) distinct stages of SOA maturity are identified: "Early Learning", "Integration", "Re-Engineering" and "Maturity".

In the "Early Learning" stage, basic capabilities are put in place as experimental activity. Short-term ROI and specific project based activities make this stage a low risk and impact stage.

During the "Integration" stage, an organization is likely to see opportunities from integration with existing core systems. In addition, some organizations may choose to enable key services which can provide immediate business benefit, whereas others will choose to service-enable a complete set of common services.

In summary, in the "Integration" stage, basic business and technical services are becoming available, together with the service infrastructure.

Once the SOA infrastructure is established and the basic services are available to be "consumed", the "Re-Engineering" stage can take place. In this stage, many enterprises can see opportunity in re-engineering business models. Also at this

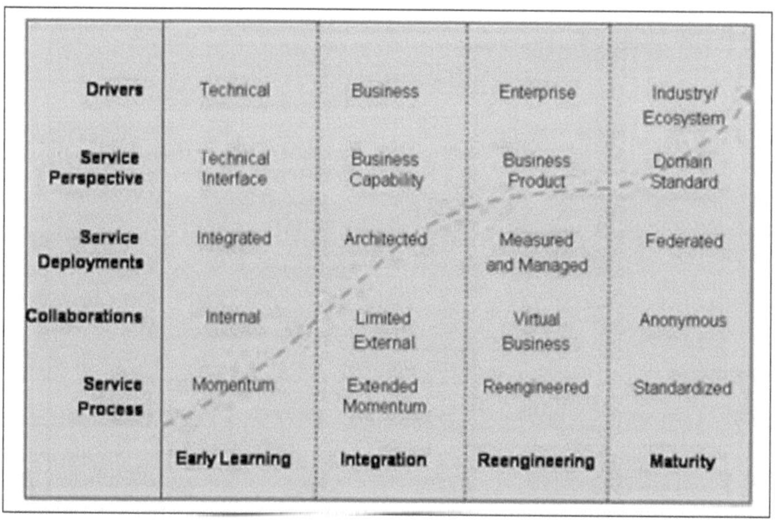

Fig. I.1 A Generic SOA Roadmap, According To CBDI Forum—[*Source* (CBDI 2006)]

stage it is likely that the increasing business criticality will require more sophisticated management techniques to be employed in order to monitor and manage service provision.

The final stage of the SOA Roadmap, according to CBDI (2006), is the "Maturity" stage. Relative maturity of service architecture can be realized at this stage, if most (of an enterprise) business products and processes are aligned with the service perspective. In this mature state, there is seamless integration throughout the enterprise systems, its customers' and suppliers' systems, and there is a high level of automation of the entire service life cycle. In addition, there is an integrated management of business process execution through to deployed resources.

IBM's SOA Maturity Model

IBM (2006), another big player in this arena -SOA-, proposes a SOA maturity model and a process for achieving desirable stages of maturity called the Service Integration Maturity Model (SIMM) as shown in Fig. I.2. The level of service decoupling and amount of flexibility achievable at each stage of maturity are what make up the following seven levels of maturity:

(a) Silo (data integration);
(b) Integrated (application integration);
(c) Componentized (functional integration);
(d) Simple services (process integration);

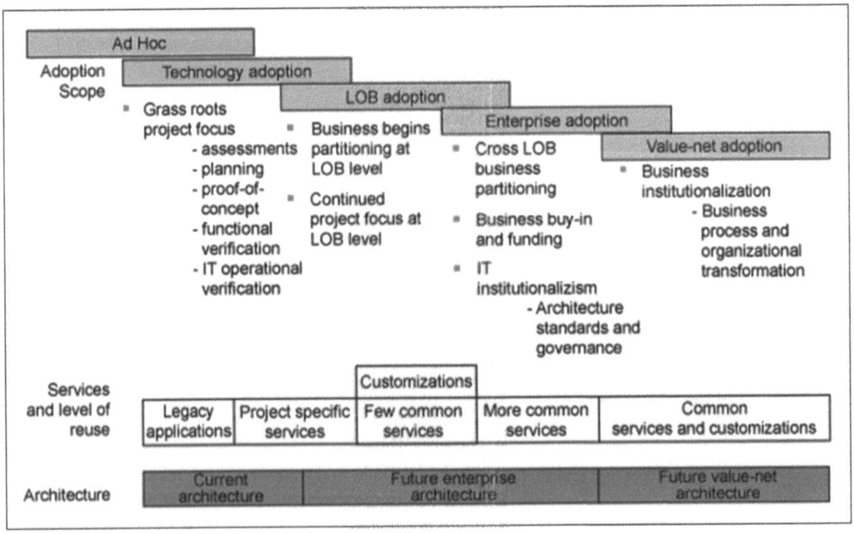

Fig. I.2 IBM's SOA Maturity Model—[*Source* (IBM 2006)]

(e) Composite services (supply-chain integration);
(f) Virtualized services (virtual infrastructure);
(g) Dynamically reconfigurable services (eco-system integration)

Each of these levels has a detailed set of characteristics and criteria for assessment, and what follows is a brief description of the highlights of each level:

(a) *Level One*: The organization starts from proprietary and quite ad-hoc integration;
(b) *Level Two*: The organization moves toward some form of Enterprise Application Integration (EAI), with proprietary connections and integration points;
(c) *Level Three*: At this level, the organization "component-izes" and modularizes major or critical parts of its application portfolio. In particular, "it uses legacy transformation and renovation methods to re-factor legacy J2EE or .NET-based systems with clear component boundaries and scope, exposing functionality in a more modular fashion. The integration between components is through their interfaces and the contracts between them;" (IBM 2006)
(d) *Level Four*: The organization embarks on the early phases of SOA by defining and exposing services for consumption internally or externally for business partners;
(e) *Level Five*: In this level, "the organization extends its influence into the value chain and into the service eco-system. Services form a contract among suppliers, consumers, and brokers who can build their own eco-system for on-demand interaction;" (IBM 2006)
(f) *Level Six*: According to IBM's Maturity Model, in this level "the organization now creates a virtualized infrastructure to run applications. It achieves this level after decoupling the application, its services, components, and flows. Now the infrastructure is more finely tuned, and the notions of the grid and the grid service render it more agile. It externalizes its monitoring, management, and events;" (IBM 2006)
(g) *Level Seven*: The organization now has a dynamically re-configurable software architecture. Particularly, "the organization can compose services at run-time using externalized policy descriptions, management, and monitoring." (IBM 2006)

BEA's SOA Maturity Model

BEA Systems (2006), a leader in the areas of middleware infrastructure and SOA, came up with a SOA Assessment Service, an approach for creating a roadmap to SOA for organizations, which provides an analysis of the business needs and IT constraints that can be addressed through SOA. BEA's SOA Assessment Service provides an overview of SOA methodologies and underlying technologies, and an understanding of the organizational challenges in moving towards SOA and a "reusability" driven IT culture. It also helps organizations by producing a Gap Analysis between the current and desired SOA state.

Accenture's SOA Maturity Model

Another big player in the area of SOA implementations (and thus a provider of SOA Readiness Assessment and Maturity tools) is Accenture (2006), one of the biggest IT Consulting companies. Accenture's SOA Assessment Model is a tool that provides customized, practical analysis to support an organization's SOA planning, and to help it move more rapidly toward implementation.

In particular, Accenture's SOA Assessment Model is comprised of a Web-based diagnostic and prescriptive tool which provides unique value by:

(b) Examining an organization's existing and planned business strategies to identify key areas where SOA can have a significant positive impact;
(c) Providing a functional and technical high-level Gap Analysis (between the current and the desired SOA state);
(d) Suggesting the next steps on how to bridge the gaps (as identified by the Gap Analysis) and identifying accelerators to ensure that an organization's SOA adoption is successful and that it produces the intended business benefits;
(e) Recommending the applicable SOA composite design pattern, along with the various foundational design patterns, to help an enterprise conceptualize its solution;
(f) Providing a case study detailing how an implementation has produced a dramatic return on investment (ROI).

Sun Microsystems' SOA Maturity Model

Sun Microsystems (Sun 2006), another big player in the area of SOA solutions, also proposed a SOA Readiness Assessment Service which involves four major activities:

(a) Discovery. During this activity a determination of the existing business and technology strategies, methodologies, processes, infrastructure, and enterprise application integration (EAI) and business-to-business (B2B) architectures through interviews and document collection, takes place;
(b) Analysis. In the Analysis activity, a comparison of the existing SOA state and the Sun's SOA guidelines occurs in order to determine the best practices and gaps;
(c) Working Sessions. This is the activity where Sun's IT architects and consultants conduct multiple 1–2 day architecture working sessions with key customer stakeholders to identify SOA opportunities and challenges, explain findings, and validate tactical and strategic recommendations;
(d) Findings. During this activity, a delivery of SOA Readiness Assessment results in the form of a written report and/or onsite presentation to key customer stakeholders. The SOA Readiness Assessment will include both tactical and strategic recommendations for migration to a SOA.

HP's and Oracle's SOA Maturity Model

Even though the above-mentioned SOA Readiness Assessment tools and SOA Maturity Models are relatively complete, the most influential to our proposed Maturity Model is the one defined by HP (HP 2006) and Oracle (Oracle 2006), as shown in Fig. I.3 below. This is because of its clear identification of key indicators that help determine enterprise's current SOA maturity level and how this could be leveraged.

According to these vendors' SOA maturity model, assessment of an SOA program is based on a core set of domains (that comprise the SOA Domain Model). The HP's (2006) and Oracle's (2006) SOA Domain Model provides a unifying framework for the SOA Readiness Assessment and it also forms the basis for their proposed Maturity Model.

The SOA Domain Model proposed by HP and Oracle are based on the analysis of the many different SOA implementations the companies' consultants have undertaken and eight primary domains were identified:

(a) Business; According to HP's and Oracle's joint effort on SOA, its adoption has an impact upon and provides benefits for both business and IT. In order to successfully adopt SOA across the enterprise, it's imperative that both business and IT commit to the program. Both sides must recognize that there will be different ways of working, and both should recognize that there will be benefits realized for each.

(b) People; At the heart of the People Domain is communication. Looking at the other domains, it is clear that all the discoveries and decisions need to be communicated across the organization. People need to understand what is happening from the beginning, the goals of the SOA adoption program, how all these disparate pieces actually fit together, and what benefits will be realized—not only for the organization as a whole, but for the people specifically.

This model's providers (HP and Oracle) argue that when the SOA is adopted by an organization, people might take on new roles that are more of a combination of business and IT rather than being just one or the other. Moreover, organizational structure changes might be observed, moving away from layers of technology to structures that are more based around services. For example, rather than having a team that looks after the network or a specific software application, there will be a team that looks after the financial services, the team that overlooks inventory management services, and so on.

"Fundamentally, SOA requires a change in the way that IT people work together and in the way they work with others in the organization. Trust becomes a key issue, particularly as resources are shared across an enterprise and business managers find that a system that was once solely under their control is now being shared with others. Similarly, they will find that they are now dependent on other groups and other parts of the business to meet their needs." (HP 2006, Oracle 2006).

(c) Program Management; In the Program Management Domain an important element is the organizational span of the SOA rollout across teams,

LEVELS / DOMAINS	Level1 (Ad-hoc)	Level2 (Basic)	Level3 (Standardized)	Level4 (Managed)	Level5 (Adaptive)
Business	Minimal interest in SOA	Aware of SOA	Generally complies with SOA	Proactively supports SOA	SOA fundamental to business ops
Program Management	SOA project focused	SOA efforts are BU-based	SOA federated, not integrated	SOA integrated at corporate level	SOA enterprise wide, extends to partners
Governance	Some acknowledgeme nt of issues	Some processes, individual responsibility	Guidelines defined and integrated	Value of governance understood	Advanced understanding
Architecture	Limited or ineffective architecture	Program exists & architecture well defined	All IT initiatives comply	Business driven and auditable linked	Architecture and business executed as integrated
Operations and Management	No mgmt. of services, infrastructures elements only	Mgmt. in terms of Slaps	Mgmt. of business services	Proactive mgmt. of svcs. linked to component svcs.	Integrated mgmt. of business services into operations
Supply and Demand	Business needs met using tech. components	Services provided internally	Value-based sourcing	Sourced from multiple providers	Dynamic service sourcing of multiple sources
People	Little or no knowledge of SOA	Knowledge limited to IT mgmt. & architects	SOA education required for IT staff	Ongoing education for all staff	SOA is embraced & actively promoted
Enabling Technologies	No service Infrastructure	SOA infrastructure limited	Standard enterprise-wide SOA	Large-scale managed SOA infrastructure	Integrated, dynamic SOA infrastructure

Fig. I.3 HP's and Oracle's SOA Maturity Model—[*Source* (HP 2006, Oracle 2006)]

departments, business units, and the entire enterprise, as well as managing the depth of the service portfolio.In particular, the two providers propose that "the SOA adoption requires an iterative approach, with SOA rolled out as a series of steps. Each step provides a complete business solution, and each step delivers measurable business value." (HP 2006, Oracle 2006).

(d) Governance; This domain concerns the models, systems and processes by which an enterprise's operations are governed: "What are the key activities that need to happen?", "What are the key decisions that need to be made?", "Who is responsible for making those decisions?", "What information is needed to be able to make those decisions?", etc.The Governance Domain must include things such as portfolio management, risk and compliance management and, of course, financial management. While most of the focus is upon governance of information technology, it is imperative that the IT governance models are coupled directly to the enterprise's overall corporate governance model.

(e) Architecture; The Architecture Domain covers the full architecture spectrum: enterprise architecture, solution architecture and technology architecture.In an SOA world, with well-defined, loosely coupled, modular services, such an architectural model allows the rapid and cost-effective response to change and the capability to take advantage of opportunities, with both confidence and predictable results.

(f) Enabling Technologies; "The Enabling Technology Domain encompasses the tools and technologies needed to support achievement of the goals of the other seven SOA domains and to realize the infrastructure needed to support a SOA within an enterprise." (HP 2006, Oracle 2006).

(g) Operations and Management; In this domain "processes and policies defined in the governance model are applied. The domain covers all aspects of operating and managing the SOA." (HP 2006, Oracle 2006).

(h) Supply and Demand; The Supply and Demand Domain is the one where we anticipate significant growth and increasing focus as an enterprise's SOA implementation matures.

HP (2006) and Oracle (2006) argue that when all these domains are mastered, an enterprise does have the capabilities and the assets in place to effectively adopt and operate an SOA.

The HP's and Oracle's SOA Maturity Model is structured as a matrix with two primary axes (Fig. I.3). Each row represents one of the eight SOA domains. Each column represents a different level of maturity using five maturity levels, from the least mature in the left-most column to the most mature in the right-hand column:

Level 1 (Ad-hoc); Level 1 is the starting point for most SOA "journeys". For an enterprise at Level 1, SOA is a relatively new concept. The enterprise has either taken no real steps toward SOA, or they may have conducted some limited, initial Web Services or Service-based activities that are project-centric, experimental and often technology-focused.

Level 2 (Basic); Typically, enterprises that are at Level 2 maturity have made a firm commitment to adopting SOA, although this may still be limited to certain parts of the organization. They will have completed a pilot or initial project with SOA applied consistently across the project and will have deployed a set of services that are in production use by the enterprise.

Level 3 (Standardized); Enterprises that have achieved Level 3 maturity have adopted SOA as a strategic enterprise-wide architectural principle. An enterprise service catalog has been established and an enterprise-wide service model is defined and used. A set of SOA standards has been defined and is applied across the enterprise. The enterprise has governance systems that ensure that all new projects are compliant with the enterprise's SOA principles.

Level 4 (Managed); At Level 4, SOA is fundamental to the way the enterprise operates both its business and its information technology, and services may extend outside the enterprise. The enterprise's service portfolio is well-managed with quantitative, integrated, enterprise-wide visibility and control of service operations. Service operational metrics are collected and reported in both business and technology contexts according to the audience.

Level 5 (Adaptive); When an enterprise reaches Level 5 maturity, it truly can be described as an Adaptive Enterprise. The whole enterprise operates a dynamic SOA with business and IT synchronized to achieve an optimum balance of agility, performance, risk and cost.

References

Accenture (2006). Retrieved December, 2006, from http://www.accenture.com/
BEA Systems (2006). Retrieved December, 2006, from http://www.cio.com
CBDI (2006). Retrieved December, 2006, from http://www.cbdiforum.com/
HP (2006). Retrieved December, 2006, from http://www.hp.com
IBM (2006). Retrieved December, 2006, from http://www.ibm.com
ORACLE (2006). Retrieved December, 2006, from http://www.oracle.com
SUN (2006). Retrieved December, 2006, from http://www.sun.com

MIX
Papier aus verantwortungsvollen Quellen
Paper from responsible sources
FSC® C105338

If you have any concerns about our products,
you can contact us on
ProductSafety@springernature.com

In case Publisher is established outside the EU,
the EU authorized representative is:
Springer Nature Customer Service Center GmbH
Europaplatz 3, 69115 Heidelberg, Germany

Printed by Libri Plureos GmbH
in Hamburg, Germany